T0259715

Textile Science and Clothing Technology

Series editor

Subramanian Senthilkannan Muthu, SGS Hong Kong Limited, Hong Kong, Hong Kong

More information about this series at http://www.springer.com/series/13111

Subramanian Senthilkannan Muthu
Editor

Textiles and Clothing Sustainability

Nanotextiles and Sustainability

Editor
Subramanian Senthilkannan Muthu
Environmental Services Manager-Asia
SGS Hong Kong Limited
Hong Kong
Hong Kong

ISSN 2197-9863 ISSN 2197-9871 (electronic)
Textile Science and Clothing Technology
ISBN 978-981-10-9554-2 ISBN 978-981-10-2188-6 (eBook)
DOI 10.1007/978-981-10-2188-6

Printed on acid-free paper

This Springer imprint is published by Springer Nature
The registered company is Springer Science+Business Media Singapore Pte Ltd.

Contents

Advances in Nanotextile Finishes—An Approach Towards Sustainability

N. Gokarneshan, P.T. Chandrasekar and L. Suvitha

Abstract The chapter critically surveys the recent development trends in nanotextile finishes. Garments for special needs comprising of the functional aspects such as protective, medical treatment and care, have been considered through treatment with silver nanoparticles and have been related to sustainability. The micro- and nanoencapsulation of 100 % cotton denim fabric using three herbal extracts have been studied for antimicrobial efficiency, resulting in improvement in durability and good resistance to microbes over 30 industrial washes. The synthesis, characterization, and application of nanochitosan on cotton fabric has been studied, and the treated fabrics were evaluated for appearance, tensile, absorbency, stiffness, dyeing behaviour, wrinkle recovery, and antibacterial properties. Polyester fabric has been treated with nanosized dispersed dye particles without carrier, using ultrasound. This has been used for optimizing the parameters for the preparation of the printing paste. Attempt has been made to improve the handle property of jute polyester-blended yarn to produce union fabric with cotton yarn, intended for winter garment. The findings indicate that nano–micropolysiloxane-based finishing exhibit better improvement in the surface morphology, handling, and recovery property of the fabric as compared with other finishing combinations. Viscose fabrics have been modified to improve the attraction for metal oxides such as aluminium, zinc, or titanium in order to impart antimicrobial activity against two types of microorganisms. Nanosafe textile using the extracts of yellow papaya peel has been developed by extracellular synthesis of highly stable silver nanoparticles. Cotton fabrics with smart properties have been developed by functional finishing with stimuli-responsive nanogel using a combination of biopolymer and synthetic

N. Gokarneshan (✉) · P.T. Chandrasekar · L. Suvitha
Department of Textile Technology, Park College of Engineering and Technology, Coimbatore, Tamil Nadu, India
e-mail: advaitcbe@rediffmail.com

S.S. Muthu (ed.), *Textiles and Clothing Sustainability*,
Textile Science and Clothing Technology,
DOI 10.1007/978-981-10-2188-6_1

polymer in the synthesis of nanogel. The findings reveal that the application of nanogel as a smart finishing system affects the intrinsic cotton properties and also improves the common textile quality by providing new features of stimuli responsiveness. Above all these, the newer concepts of the nanotextile finishes discussed in the chapter promise enhancement in the existing properties of textile materials, increase durability, promote ecofriendliness and economy, and thus pave the way towards achieving better sustainability.

Keywords Chitosan · Disperse dye · Viscose · Nanogel · Silver particles · Nanoencapsulation · Nanopolysiloxane

1 Introduction

The new millennium has witnessed significant research works around the globe in the area of nanotextile finishing (Gokarneshan et al. 2013). A number of innovative approaches have been attempted during the recent years and tend to exploit the functional properties and tailor the fabrics to suit specific requirements. Some of the efforts have been directed towards improvement in antibacterial efficacy of the fabrics so treated (Gokarneshan et al. 2012). When considering garments for special needs, three aspects have been considered, namely, protective, treatment, and caring. Efforts have been directed to improve the comfort and functionality of these groups clothing with nanotechnology by assessing all of these concerns and comparing the benefits of nanotechnology with its disadvantages (Ebrahim and Mansour 2013). Application of antimicrobial finishes using natural materials has been the popular trend, which promotes natural and ecofriendly lifestyle. Attempt has been made through the use of plant extracts containing active substances on fabrics, so as to maker them microbial resistant (Sumithra and Vasugi Raaja 2012). The textile wet processing operations are going through a stage of green revolution, wherein many types of conventional and non-eco-friendly chemicals are being replaced by natural-based products that are safe to environment and health during manufacturing and usage. The synthesis and characterization of nanochitosan on cotton fabric has been studied. The nanochitosan-treated fabrics are then tested for appearance, tensile, absorbency, stiffness, dyeing behaviour, wrinkle recovery, and antibacterial properties (Chattopadhyay and Inamdar 2013). In an interesting study, a nanoparticle-sized disperse dye treated with ultrasound has been applied on polyester fabric without using a carrier. The dyeing and process parameters considered include *K*/*S* values, dye particle size, dye exposure to ultrasound waves, printing paste pH, steaming conditions of prints, morphological study using SEM and TEM of dye particles, and fastness properties of the prints (Osman and Khairy 2013). Even though jute fibre has unique properties such as roughness, coarseness, and stiffness, it can cause problems during fabric formation and performance of final product as well, rendering it unsuitable for apparel. Hence efforts have been taken to improve the handle property of jute fabric through application of nanopolysiloxane-based finishing both in individual as well as in combination with other finishing chemicals by conventional pad-dry-cure method (Lakshmanan et al. 2014). Textile fabrics are generally subjected to repeated laundering during their lifetime, and hence the washing durability of nanometal-treated fabric is of significant importance. In this context, attempt has been made to prepare permanent antimicrobial viscose fabrics by fixation of propionic acid groups at lower temperature (below 100 °C), as active centres, onto the cellulosic polymeric chain. The added carboxylic groups are believed to act as favourable centres for some oxides such as titanium oxide, zinc oxide, or aluminium oxide nanoparticles. The efficiency of the antimicrobial activity, considering the permanent performance against selected microorganisms onto modified textile, has also been evaluated (El-Sayeed et al. 2015). Cotton fabrics have been functionally finished with stimuli-responsive nanogel comprising of biopolymer chitosan and a synthetic polymer so as to obtain smart properties (Bashari et al. 2015). The above approaches enable a more effective way of nanofinishing of textile materials and hold the prospects of greater viability coupled with sustainability, since they address the concerns related to the field of application of nanotechnology in textiles. The chapter focuses on the advantages of the newer methods of nanotextile finishing which hold the prospects of wider areas of applications.

2 Treatment of Nanomaterials on Garments for Special Needs

2.1 Overview

The mechanical and physical properties of cotton fabrics can be modified to meet special needs garments, by application of nanosilver finishes. Accordingly fabrics have been subjected to the nanosilver finishing process with 5 different solution concentrations of 100, 200, 300, 400 and 500 ppm. It has been found that nanosilver finishing minimized the air and water vapour permeability of the fabric, which could be due to the nanosilver particles that fill the fabric pores. The wrinkle recovery angle of the nanofinished fabrics has decreased in both directions due to the generation of links on the fabric by nanosilver particles (Ebrahim and Mansour 2013). Yarn swelling phenomena happens during the finishing process in cotton fabrics which means increment of occupied space by fibres and yarns and thus rise in the fabric thickness; owing to integration of fibres and yarns by the nanosilver particles, the breaking elongations of fabrics in both warp and weft directions were inclined to applying the nanosilver finishing process, and applying the nanosilver finishing process led to formation of links on the fabric, and consequently bending rigidity of the fabrics was increased in both warp and weft directions. Finally, it has been observed that there has been a decrease in most of the physical and chemical properties of cotton fabrics by increasing the concentration of the solution to 500 ppm.

2.2 Related Aspects

Despite the fact that the application of nanoparticles to textile materials is gaining popularity owing to their novel physicochemical properties and their potential applications, some of them are toxic or ineffective and renders them unsuitable for applications in medicine, filters, and textiles and for the exclusion of pollution. To quote some examples

(a) Improving the water-repellent property of the fabric by creating nanowhiskers (hydrocarbons which are 1/1000 size of typical cotton fibres) on the fabric
(b) Creating antistatic properties of textile which can be provided by TiO_2, ZnO, antimony-doped tin oxide (ATO), and silane nanosol, increasing the surface energy and thereby providing a very high particle retention to filters by the usage of nanofibrous webs on them
(c) Using nanotitanium dioxide and nanosilica to advance the wrinkle resistance of cotton and silk respectively, employing nanosized TiO_2 and ZnO in order to absorb and scatter UV radiation more effectively regarding the larger surface area and blocking ability of so-called particles
(d) Covering the cotton fibres in a fuzz of minute whiskers and creating fewer points of contact of dirt, thus the fabric has been rendered super-hydrophobic, and the self-cleaning property can be developed in this way.

Eventually, antibacterial properties can be imparted by using nanosized silver, titanium dioxide, and zinc oxide (Wong et al. 2006a; Anna et al. 2007; Parthasarathi 2008). The nanosilver particles are given a greater focus, as they have an extremely large relative surface area, and so their contact with bacteria or fungi is increased, thereby resulting in great improvement in their bactericidal and fungicidal effectiveness.

Cotton is generally used in the production of textiles for sport and leisure activities owing to its outstanding moisture absorption ability. Since cotton is moist, it is susceptible to bacterial attack. There is a characteristic odour in decomposed products of body secretions (Gorensek and Recel 2007). In recent research, a good antibacterial effect of nanosized silver colloidal solution on polymer and textile fabrics was shown (Lee and Jeong 2005; Yeo et al. 2003). It has been intended to not only study the synthesis of silver nanoparticles used in microwave radiation as a heating source but also investigate the role of nanotechnology in improving sustainability cotton fabrics and the relationship between the physical and mechanical properties and the content of nanosized silver on cotton fabric. Generally, this utilization can be categorized into two main areas: firstly, application of nanofibres and, secondly, application of nanoparticles in different domains. Here some of these applications can be named. Some examples are the use of polymeric nanofibres and their composites in drug delivery systems, tissue engineering, reinforcement of some composites, transistors, capacitors, and so on (Hatiboglu 2006; Huanga et al. 2003; Tan and Lim 2006).

2.3 *Technical Details*

100 % plain woven cotton fabrics have been used. For investigating the influence of nanosilver finishes on the physical and mechanical properties of the fabric, six different fabrics have been tested (Ebrahim and Mansour 2013). Finished fabrics have been compared with unfinished fabrics. The finishing has been done with five different solution concentrations, so as to identify a trend in changes. Fabrics have been soaked in 50 °C suspensions with five distinct concentrations of nanosilver particles for 30 min and then dried in the open air (known as 'exhausting finishing'). The physical and mechanical properties including air permeability **(A.P)**, wrinkle recovery **(W.R)**, water vapour permeability **(W.V.P)**, thickness **(T)**, breaking strength **(B.S)**, breaking elongation **(B.E)**, and bending rigidity **(B.R)** have been measured as a testing procedure (Ansari and Maleki 2007).

3 Physical and Mechanical Properties

3.1 *Air Permeability*

The test results of air permeability are shown in Fig. 1. A severe drop in the air permeability property can be clearly observed by applying nanosilver finish on the fabric, and also a gradual fall by increasing the solution concentration (Ebrahim and Mansour 2013) is observed. This occurs due to the reduced fabric pores, which have been padded by nanosilver particles. As the nanoparticles are very small, increasing the concentration does not significantly influence the air permeability property.

3.2 *Wrinkle Recovery*

The test results of wrinkle recovery in the warp and weft directions are depicted in Figs. 2 and 3. There is a high angle of recovery in the warp direction in sample A,

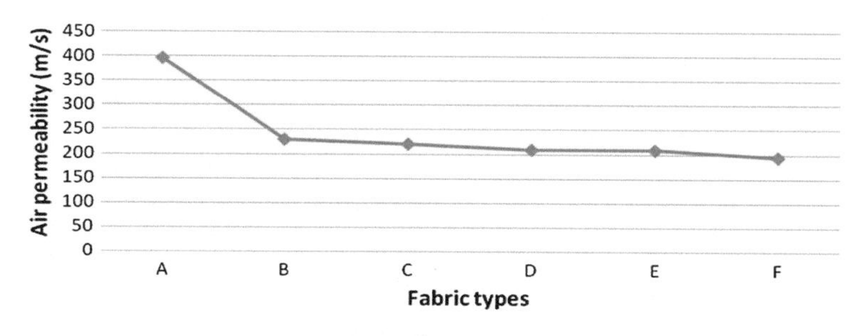

Fig. 1 Air permeability of fabrics (Ebrahim and Mansour 2013)

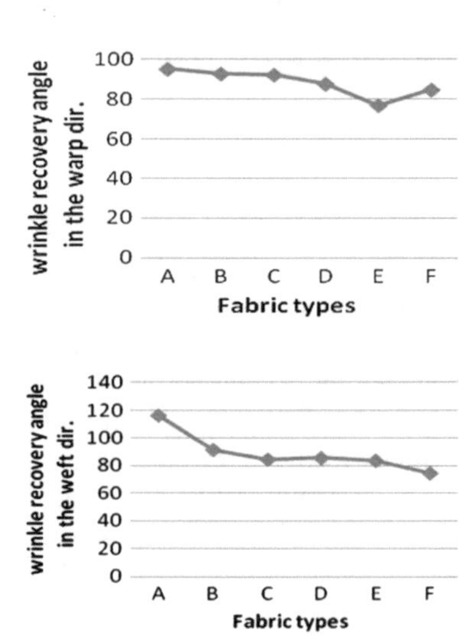

Fig. 2 Wrinkle recovery angle of fabrics in weft direction (Ebrahim and Mansour 2013)

Fig. 3 Wrinkle recovery angle of fabrics in warp direction (Ebrahim and Mansour 2013)

and it goes down gradually in sample E and then rises again in the sample F. A downward trend in wrinkle recovery is observed due to the links which have been created by nanosilver particles on the fabric, but the created links decrease upon increasing solution concentration to 500 ppm. In the case of Fabric A, the recovery angle in the weft direction is high, and there has been a moderate drop in this value for the other samples. However, no regular trend is seen, and it is due to the unevenness in finishing (Ebrahim and Mansour 2013).

3.3 Water Vapour Permeability (WVP)

The water vapour permeability percentage of the fabrics is depicted in Fig. 4. It can be generally concluded that WVP% of nanofinished fabrics are lower than the unfinished one. The explanation relating to air permeability holds valid in this case too. The highest value of WVP% is found in Fabric A, while the lowest is in Fabric F. The irregularity arises due to the unevenness of finishing process (Ebrahim and Mansour 2013).

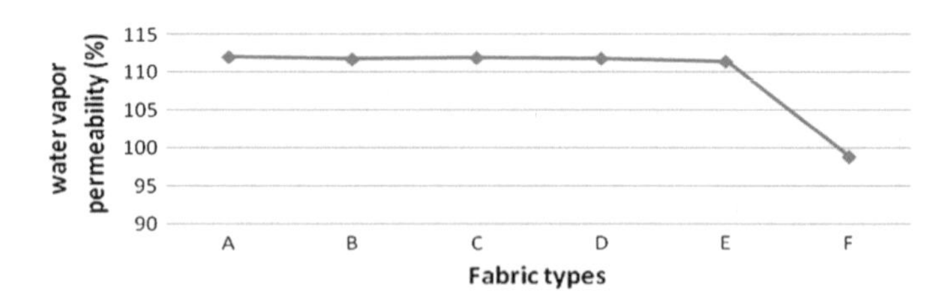

Fig. 4 Water vapour permeability of fabrics (Ebrahim and Mansour 2013)

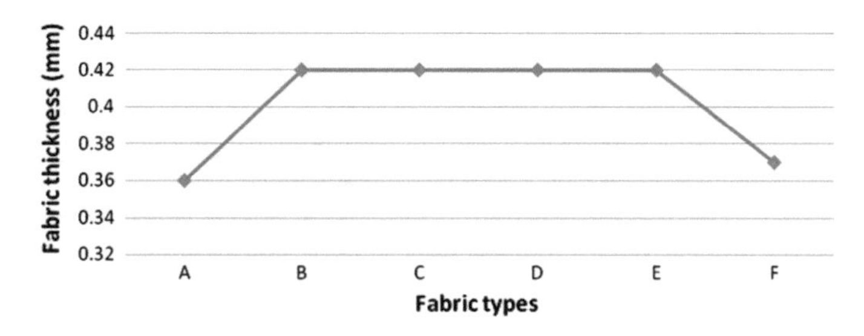

Fig. 5 Thickness of fabrics (Ebrahim and Mansour 2013)

3.4 *Thickness*

Figure 5 depicts the thickness values of the fabrics. It is evident that the thickness values of nanofinished fabrics are more than those of the unfinished fabric. The higher thickness values of the nanofinished fabrics are due to the yarn swelling phenomena that occur during the finishing process (Ebrahim and Mansour 2013). However, increasing solution concentration to 500 ppm resulted in a severe fall, which occurs due to decreasing of yarn swelling.

4 Influence of Nanosilver Finish on Mechanical Properties

4.1 *Breaking Strength*

Figure 6 shows the breaking strength of the fabrics in the warp direction. The lowest breaking strength has been exhibited by Fabric F and the highest by Fabric C. The general trend seen is that the breaking strength of nanofinished fabrics is more than unfinished fabric due to the linkage formation between fibres and yarns. However, the exact trend is not predictable. The breaking strength of the fabrics in the weft direction is shown in Fig. 7 (Ebrahim and Mansour 2013). Fabric D exhibits the highest breaking strength. However, there is no significant difference between other fabrics.

4.2 *Breaking Elongation*

Figures 8 and 9 typify the breaking elongations of the fabrics in both warp and weft directions. Significant differences can be seen in the breaking elongation of fabrics between nanofinished and unfinished fabrics. The necessary difference arises due to consolidation of fibres and yarns by the nanosilver particles (Ebrahim and Mansour 2013).

Fig. 6 Breaking strength of fabrics in weft direction (Ebrahim and Mansour 2013)

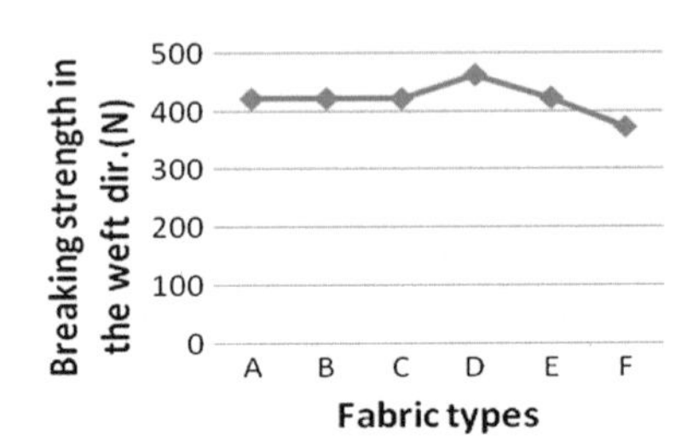

Fig. 7 Breaking strength of fabrics in warp direction (Ebrahim and Mansour 2013)

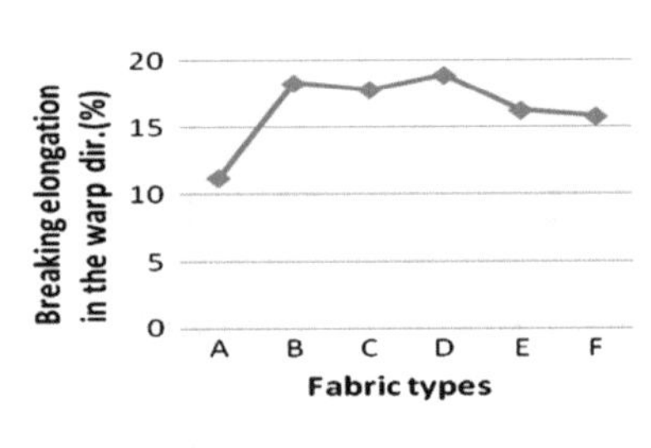

Fig. 8 Breaking elongation of fabrics in warp direction (Ebrahim and Mansour 2013)

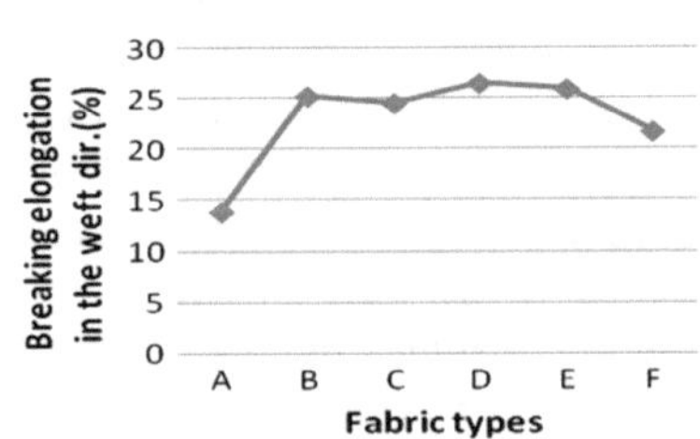

Fig. 9 Breaking elongation of fabrics in weft direction (Ebrahim and Mansour 2013)

4.3 Bending Rigidity

Figures 10 and 11 show the bending rigidity values of fabrics in the warp and weft directions, respectively. Fabric A has bending rigidity of 71.2 (mgr cm) in the warp direction and 72.5 (mgr cm) in the weft direction. In the case of Fabric E the bending rigidity values started to increase, and they reached to 126 (mgr cm) in the warp direction and 124 (mgr cm) in the weft direction (Ebrahim and Mansour 2013). This gradual upward trend is caused by the links which was formed on the yarns and

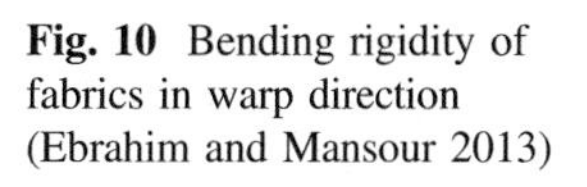

Fig. 10 Bending rigidity of fabrics in warp direction (Ebrahim and Mansour 2013)

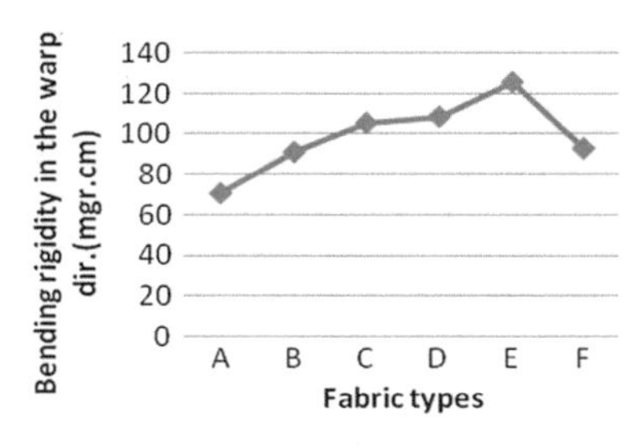

Fig. 11 Bending rigidity of fabrics in weft direction (Ebrahim and Mansour 2013)

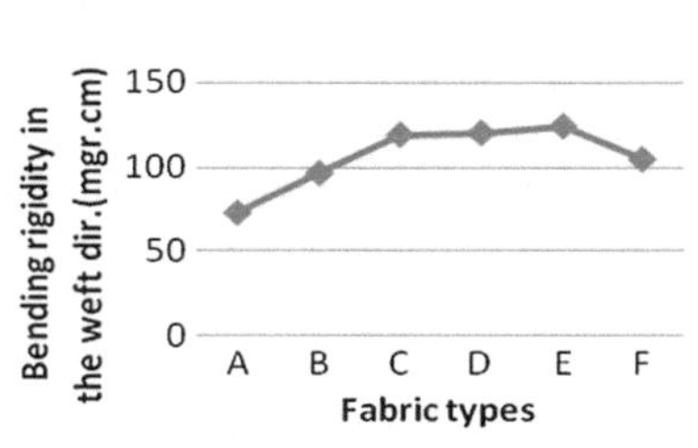

fabrics by nanosilver finishing. In the case of Fabric F, an increase in solution concentration to 500 ppm leads to a gradual fall, owing to decreasing form of links.

5 Micro- and Nanoencapsulation of Denim Fabrics

5.1 Overview

The use of herbal extracts (*Ricinus communis, Senna auriculata* and *Euphorbia hirta*) holds good prospects as antimicrobial finishing on denim fabric. The finished fabric exhibits maximum antibacterial activity against both *Escherichia coli* and *Staphylococcus aureus*. The microencapsulation and nanoencapsulation techniques increase the durability of the herbal finishing. The findings reveal antimicrobial effect even for the washed fabrics against the standard strains (Sumithra and Vasugi Raaja 2012).

5.2 Related Aspects

It is a well known fact that bacteria can grow and survive on fabrics commonly used in healthcare environments for more than 90 days and hence contribute to the transmission of diseases (Gaurav 2005; Subhash and Sarkar Ajoy 2010). The recent trend has been the use of antimicrobial finishes using natural sources, which promotes natural and ecofriendly lifestyle (Natarajan 2002). Attempt has been made through application of plant extracts containing active substances, so as to explore the possibility of making the cloth microbial resistant (Sumithra and Vasugi Raaja 2012). These natural antimicrobial substances are not only ecofriendly but are also taken from renewable sources (Gaurav 2005). Bacterial growth in textile materials leads to the deterioration of fabric properties and produces foul smells, skins irritation, and cross-infections (Subhash and Sarkar Ajoy 2010). Microbes are small organisms that cannot be seen by the naked eye and include a variety of microorganisms such as bacteria, algae, and fungi (Natarajan 2002). Encapsulation involves covering small solid, liquid, or gaseous substrates by means of a polymeric or inorganic shell. The resulting capsules or particles normally range from micrometres to millimetres in size (Holme 2002).

The microencapsulation has multivaried applications that include controlled release of the active components, particle coating, flavour stabilization, taste masking, physical/chemical stabilization, improvement in shelf life, and prevention of exposure of the active material to the surroundings (Bhoomika et al. 2007). Two common technologies can be used to obtain such nanocapsules and microcapsules, namely the interfacial polymerization of a monomer and the interfacial deposition of a preformed polymer (Shilpa 2004). Denim has become very popular due to its suitability for many occasions rather than merely complementing a rugged style, and is also used irrespective of demographic differences (Srikanth 2010). The durability of denim as only the warp yarns go through the dyeing process, while the weft yarn is left natural without undergoing any chemical process. This is the advantage of yarn-dyed fabric over piece-dyed fabric (Thies 2005). The denim wear is gaining importance tremendously each year and has also increased its worldwide market share unpredictably during the past few decades. The consumer needs have been focussed towards the latest developments and new styles, and also there is increased awareness of special finishes and process treatments given to the garment to make them ecofriendly and user-friendly. The discussion herein focuses on screening for the antimicrobial activity of the natural herbs and providing the denim fabric with the antimicrobial finish from the screened herbal extracts. AATCC methods are followed relating to the combinations and conditions. Microencapsulation and nanoencapsulation methods have been used to increase the washing durability of the antimicrobial finish provided from the herbal extracts.

5.3 Technical Details

For the application of antimicrobial finish, 100 % cotton denim fabric has been used. The antimicrobial extract has been obtained using the herbs *R. communis* (leaves), *S. auriculata* (leaves), and *E. hirta* (leaves, stem and flower). The collected plants have been dried at a temperature range of 100–140 °C, as they cannot be stored without drying, thereby preventing the breakdown of important compounds and contamination by microorganisms (Sumithra and Vasugi Raaja 2012). The fabric has been primarily washed with distilled water, air-dried, and then used for herbal finishing. Dip technique has been used to finish the fabric with herbal extract.

The fabric has been immersed in the extract for 30 min, air-dried, and then used for the antibacterial assessment using standard bacterial strains. The AATCC test method has been used, and the herbal extracts having antibacterial activity have been mixed during preliminary screening, in various combinations and conditions used for their combined antibacterial property. The denim fabric with herbal extract having the diameter of 23 mm was placed on the surface of medium and the plates were kept for incubation at 37 °C for 24 h. The zone of inhibition formed around the fabric during the end of incubation has been measured in mm and recorded. Addition of sodium alginate with subsequent spraying into the calcium chloride solution by sprayer resulted in formation of microcapsules containing herbal extract. In order to harden the capsules, the droplets have been retained in the calcium chloride solution for 15 min. The microcapsules were obtained by decantation and followed by repeated washing with isopropyl alcohol and then drying at 45 °C for 12 h. The microcapsules were applied on the fabric by exhaustion method using 8 % citric acid (cross-linking agent). The fabric was kept immersed in the solution (ML ratio 1:20) for 30 min at 50 °C in water bath. After finishing, the fabric was removed, squeezed, and dried at 80 °C in the oven for 5 min and then cured at 120 °C for 2 min. The antibacterial activity of the microcapsule-finished fabric was analysed by AATCC test method. The herbal extracts prepared were encapsulated using bovine albumin fraction as the wall material and the nanoparticles as the core material. The herbal extract-enclosed bovine serum albumin protein was prepared by coacervation process followed by cross-linking with glutaraldehyde. After glutaraldehyde treatment, for purification the solution is put in rotary vacuum evaporator to remove the organic solvent, then centrifuged at 4 °C at 10,000 rpm, suspended in 0.1 M phosphate buffer (pH 7.4), and then lyophilized with mannitol (2 % w/v). The herbal extract was incubated with the required protein solution (2 % w/v) for an hour at room temperature. The pH of the solution was adjusted to 5.5 by IMHCL using digital pH meter. Then ethanol was added to the solution in the ratio of 2:1 (v/v). The rate of ethanol addition was carefully controlled at 1 mL per min. The coacervate so formed was hardened with 25 % glutaraldehyde for 2 h to allow cross-linking of protein. Organic solvents were then removed under reduced pressure by rotary vacuum evaporator, and the resulting nanocapsules were purified by centrifugation (10,000 rpm) at 4 °C. Pellets of nanocapsules thus obtained were then suspended in phosphate buffer (pH 7.4; 0.1 M), and each sample was finally lyophilized with mannitol (2 % w/v). The nanocapsules obtained were further dried by lyophilization and then applied on the cotton fabric by exhaustion method using 8 % citric acid as binder. The fabric was finished following the conditions M/L ratio 1:20, binder (citric acid) 8 %, temperature 55 °C, and time 30 min. The AATCC test method has been used to evaluate the antibacterial activity of the nanocapsule-finished fabric. The denim fabric finished with microcapsules and nanocapsules have been assessed for their wash durability and antibacterial activity evaluated. The binding of microcapsules and alignment on to the fabric have been confirmed by scanning electron microscopy.

6 Determination of Antibacterial Activity of Finished Fabric

The antibacterial activity has been evaluated for the finished fabric using AATCC test method against *E. coli* and *S. aureus*. The zone of inhibition for the fabrics finished with methanolic extract of *R. communis, S. auriculata,* and *E. hirta* is 0 (no bacterial growth), 25, and 24 mm, respectively, for *E. coli* and 27, 30, and 29 mm, respectively, for *S. aureus*. The methanolic extract of the herbs is able to impart antibacterial efficacy to the finished fabric. Among the three herbs used, the methanolic extract of *S. auriculata* is found to give maximum antibacterial activity in comparison with the other two. The selected ratio of 1:3:2 of *R. communis, S. auriculata,* and *E. hirta*, respectively, is found to be the best combination. The fabric so treated is found to be strongly resistant to both Gram-positive and Gram-negative microorganisms. The bacterial growth is prevented by the treated fabrics. The values of the inhibition zone show that the herbal extracts prevent the bacterial growth under the fabric and also leaches out and kills the bacteria. In the case of *E. coli*, the bacterial growth rate is 2.7×10^9 cfu/mL, and for *S. aureus*, it is 2.4×10^9 cfu/mL.

Antibacterial activity of ajwain seeds extract treated fabrics using qualitative and quantitative methods has also been reported (Benita 1996). Antibacterial property of extract alone clearly shows extended zone of inhibition (20–30 mm) for both types of microorganism (Gram positive and Gram negative) when analysed by AATCC (parallel streak method). Test results show that all treated fabrics have very good antimicrobial property to both *S. aureus* and *Klebsiella pneumoniae* microorganisms. The treated fabrics do not allow the growth of bacteria under the test specimen. In all the cases, there is a good zone of inhibition ranging from 11.9 to 14.5 mm for *S. aureus* and from 6.9 to 7.1 mm for *K. pneumoniae*. It has been shown that the extracts from basil, clove, guava, jambolan, lemon balm, pomegranate, rosemary, and thyme exhibit antimicrobial activity to at least one of the tested microorganisms (Barari et al. 2009). The highest activities are seen in the extracts of clove and jambolan. In other words they are able to inhibit 9 (64.2 %) and 8 (57.1 %) kinds of microorganisms of interest, respectively.

6.1 *Microencapsulation of Herbal Extract*

The effective herbal combination is microencapsulated using ionic gelatin technique so as to increase the durability of the finished fabric. The antibacterial activity of the finished fabric and the fabric after washes have been tested and the results are presented in Table 1. It has been found that the microcapsules of the herbal extract-finished denim fabric exhibit activity for the sample after 20 washes. They exhibit potential for antimicrobial activity against *S. aureus* and *E. coli.* It is observed that the microencapsulated herbal extracts possess a very good resistance for microbes even after 15 washes. The antimicrobial efficiency has been defined in

Table 1 Antibacterial activity of the microcapsules and nanocapsule-finished fabrics (Sumithra and Vasugi Raaja 2012)

Samples	Zone of bacteriostasis (mm)		Zone of inhibition (mm)	
	Microcapsule		Nanocapsule	
	E. coli	*S. aureus*	*E. coli*	*S. aureus*
Finished fabric	24	33	30	35
Finished fabric after10 washes	0	30	29	33
Finished fabric after 20 washes	0	27	28	30
Finished fabric after 30 washes	0	0	28	25

terms of bacterial reduction percentage for directly applied and herbal microcapsules applied fabrics (Sanjay Vishwakarma et al. 2010). The herbal extracts applied both directly and microencapsulated show better activity against *S. aureus* than against *E. coli.*

Fabrics on which herbal extract has been directly applied do not exhibit much activity after 10 washes, since the extracts are coated only on the surface without any firm bonding and get removed by washing. This was ascertained when the structure of microcapsules was studied by scanning electron microscopy using image analysis technique. The antimicrobial efficacy by quantitative method in terms of bacterial reduction and the wash durability of antimicrobial activity was studied by agar diffusion method (AATCC 124). The fabrics show potential for antimicrobial activity against *S. aureus* and *E. coli.* The microencapsulated herbal extracts are highly resistant to microbes even after 15 washes. Hence, the herbal extracts are nanoencapsulated so as to increase the durability of the fabric.

6.2 Nanoencapsulation of Herbal Extract

The exhaustion method has been used to apply the nanocapsules of the herbal extracts on the denim fabric, and the antibacterial activity was evaluated for the finished fabric and for the fabrics after washes (Table 1; Fig. 12). The results reveal that the fabric finished with nanocapsules of herbal extracts is able to retain the

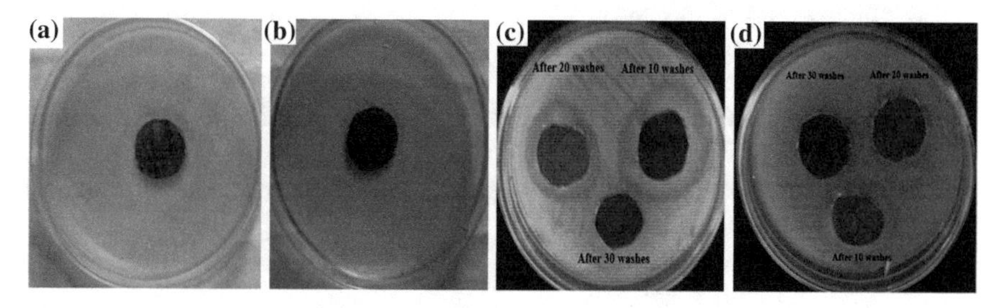

Fig. 12 Zones of inhibition of nanoencapsulated denim fabrics **a** for *E. coli*, **b** for *S. aureus*, **c** for *E. Coli* after 10, 20 and 30 washes, and **d** for *S. Aureus* after 10, 20 and 30 washes (Sumithra and Vasugi Raaja 2012)

antimicrobial activity even after 40 industrial washes, hence providing long-term durability to the finished fabric. Nanosized silver, titanium dioxide, and zinc oxide have been used for imparting antibacterial properties (Pujari et al. 2010). The effect of TiO_2 nanoparticles concentration with and without using 4 % PUA binder on antimicrobials on either screen printed or pigment-dyed silk fabric shows that the use of PUA binder alone has no effect on the antimicrobial property. But upon using TiO_2 in either printing paste or in ink preparation, good results are observed against *E. coli* and *S. aureus*. Moreover, printing yielded better results than obtained in the case of dyeing.

6.3 SEM Investigation

Figures 13 and 14 depict the SEM images at various magnification levels of nanocapsule-finished fabric and the fabrics after 10, 20, and 30 washes. It is clear that the nanocapsules not only adhere on the surface, but also penetrate into the interstices of the yarn and fabric. It shows that the nanocapsules are bonded well to the fabric surface even after 30 washes.

7 Application of Synthesized Nanochitosan on Cotton

7.1 Overview

For the preparation of nanochitosan dispersion, ionotropic gelation with penta-sodium tripolyphosphate has been used (Chattopadhyay and Inamdar 2013). The following findings accrue with regard to application of nanochitosan on cotton fabric

(a) Nanochitosan-treated cotton fabric shows reasonably good appearance and handle.
(b) There is improvement in fibre strength, which increases with the decrease in particle size of nanochitosan.
(c) There is slight reduction in elongation-to-break with the scaling down of particle size.
(d) Moisture-related property such as absorbency is affected; nevertheless, it is in acceptable limit of tolerance.
(e) The dyeability of chitosan and nanochitosan-treated cotton fabric towards direct dyes is improved significantly. The progress is sustained with reduction in particle size. The effect is further enhanced after acidification of dye bath. Fastness to washing is improved satisfactorily and fastness to rubbing slightly.
(f) Wrinkle recovery property is slightly improved and use of suitable cross-linking agent is essential.
(g) Improved antibacterial activity is observed in combined treatment of nanochitosan with nanosilver.

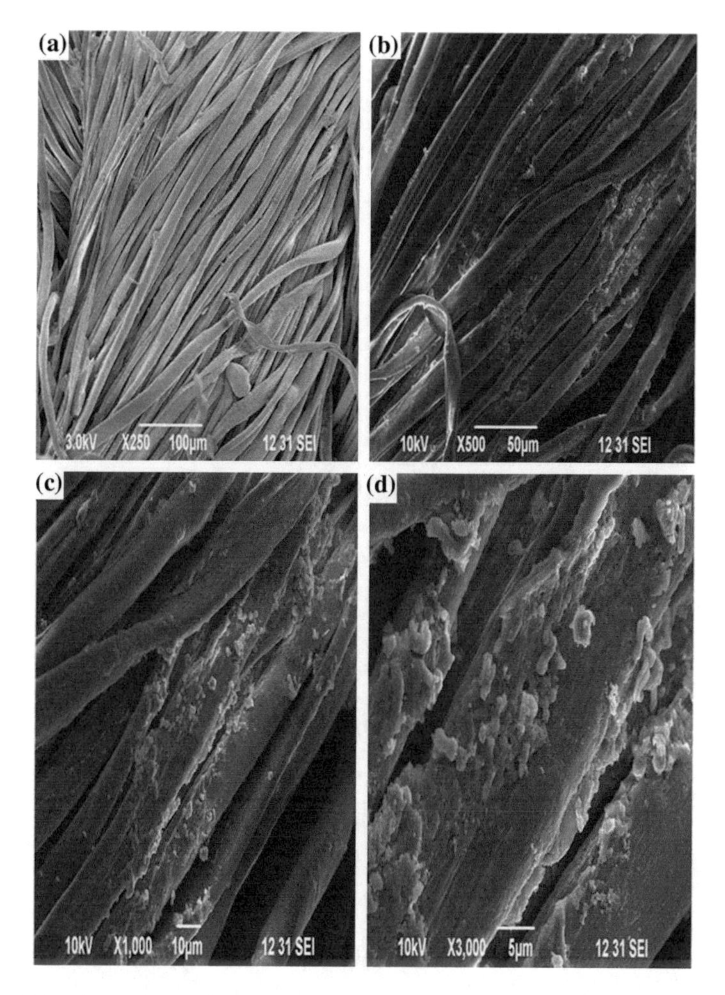

Fig. 13 SEM photographs of nanocapsules finished fabrics under different magnifications **a** ×250, **b** ×500, **c** ×1000 and **d** ×3000 (Sumithra and Vasugi Raaja 2012)

7.2 Related Aspects

The green technology is being given due attention for different wet processing operations of textiles from initial preparatory processes to final finished fabrics. Natural-based products are replacing many conventional non-ecofriendly chemicals and are considered safe to environment and health during manufacturing and usage. Applications of enzymes in preparatory and in biopolishing, natural dyes for colouration, biopolymers, and their derivatives in fibre production and finishing processes, etc., are some of them. During recent years, the biopolymer that has drawn good attention is chitosan, which is derived from alkaline deacetylation of

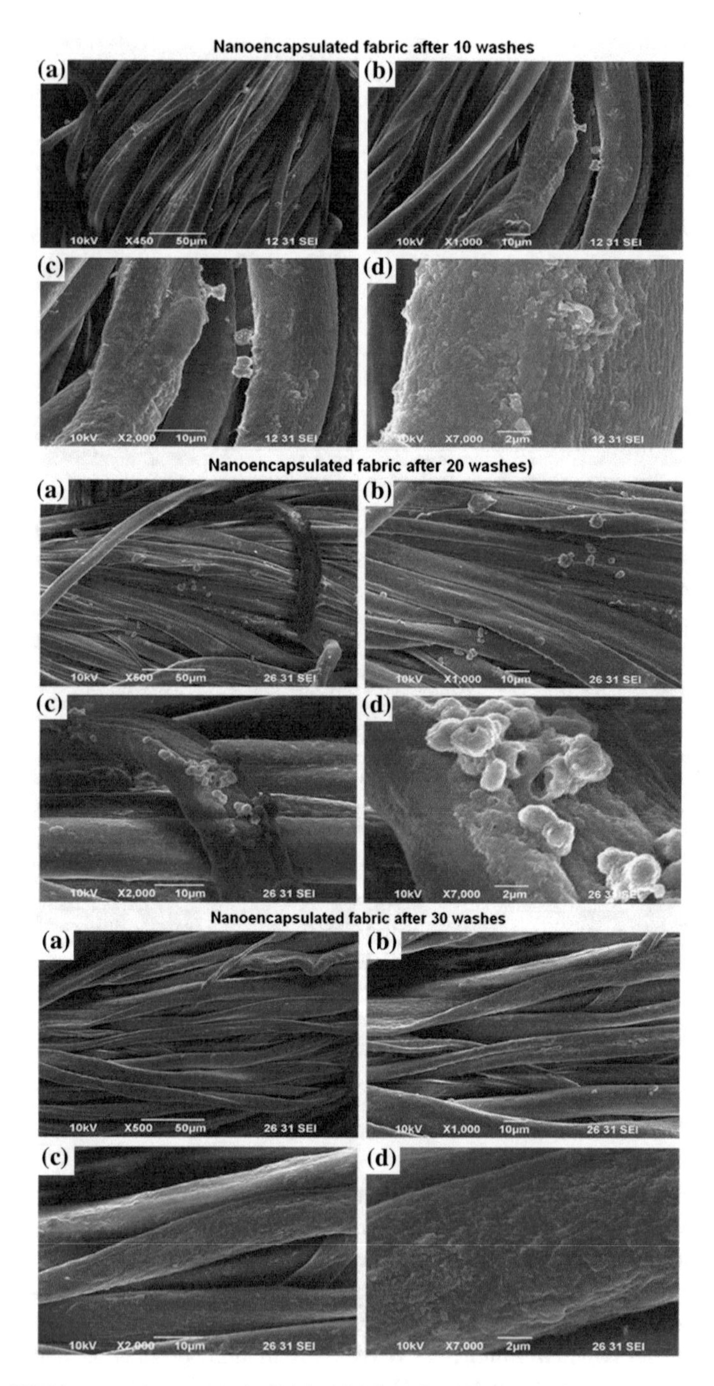

Fig. 14 SEM images of nanocapsule-finished fabrics after 10, 20, and 30 washes under different magnifications **a** ×250, **b** ×500, **c** ×1000 and **d** ×3000 (Sumithra and Vasugi Raaja 2012)

chitin (Sathianarayanan et al. 2011). The precursor chitin is a nitrogen-containing polysaccharide, which is second most abundant biopolymer after cellulose, distributed in the shells of crustaceans such as crabs, shrimps, and lobsters and in the exoskeleton of marine zoo plankton, including coral, jellyfish, and squid pens. It is totally ecofriendly and renewable (Tatiana et al. 2008; Thilagavathi et al. 2007). Chemically, chitosan is a linear linked 2-amino-2-deoxy-β-D-glucan (i.e. β-D-glucosamine) having the structure very much close to that of cellulose except the hydroxyl group in C of cellulose is being replaced by amino group in chitosan. Indeed, it is a copolymer of *N*-acetyl-glucosamine and glucosamine units. Being a primary aliphatic amine, chitosan can be protonated by various acids (El-Molla et al. 2011). In view of the many valuable inherent properties like antibacterial, antifungal, antiviral, antacid, non-toxic, totally biodegradable, biocompatible with animal and plant tissues, and film formation, fibre formation, and hydrogel formation properties, chitosan holds promise in a number of applications like biomedical, wastewater treatment, cosmetics, dentifrices, food, agriculture, pulp and paper, and textile industries (Muzzarelli 1996; Hirano 2003; No and Meyers 1995).

The application potential of chitosan in textiles has been reviewed comprehensively (Oktem 2003; Harish Prashant and Tharanathan 2007). Studies have revealed that it can also be used as a dye-fixing agent, for shade and naps coverage, for enhancing the fastness of dyed fabrics, as a binder in pigment printing, and as a thickener in printing. By virtue of its bacteria-impeding property, chitosan can prevent garments from developing bad odour (Giri Dev et al. 2005; Kean et al. 2005; Enescu 2008; Inamdar and Chattopadhyay 2006; Achwal 2000). An improved wrinkle recovery of cotton fabric is reported on finishing cotton with citric acid solution in presence of chitosan with minimum loss in tensile strength due to citric acid treatment (Achwal 2003). It is found that complete inhibition of *E. coli* and Hay bacillus bacteria is possible by treatment of cotton with 0.5 gpL chitosan concentration (Hasebe 2001). Tiwari and Gharia (Eom 2003) attempted to use chitosan as a thickener in printing paste. Performance of the prints with respect to *K/S*, wash fastness, crock fastness, and hand are observed to be unsatisfactory. Earlier studies have shown improved dyeability towards direct dyes of chitosan-pretreated cotton fabric, and the degree of improvement was found to be a function of molecular weight and concentration of chitosan (Knittel and Schollmeyer 2002). The fastness to washing of direct dyes on chitosan-pretreated fabric, however, was only slightly improved especially for the low molecular weight chitosan applications. Chitosan-treated cotton fabric also showed a substantial dyeability towards acid dyes. But, the appearance and handling of the treated fabric have been adversely affected. There has been decrease in the wrinkle recovery property. Due to the rigid film deposition of chitosan on cotton fibre surface, there is loss in inherent qualities of cotton fibres.

But it is necessary to improve above properties without changing the inherent natural qualities of cotton. It is possible by achieving the maximum penetration of polymer particles into fibre structure and increasing its effectiveness at lowest possible concentration. The decrease in particle size or reducing the viscosity of its solution helps to improve the penetration of chitosan solution. Reduction in concentration of normal chitosan in solution, however, may reduce its effectiveness and larger chain does not permit its entry into the yarn/cellulose structure. The only possible way is to reduce the particle size, which, in addition to decrease in viscosity, offers greater surface area and hence increases the effectiveness of chitosan. It forms the basis of the nanotechnology concept. The potential applications of nanochitosan are well demonstrated in medical field particularly in controlled drug delivery systems (Chattopadhyay and Inamdar 2009; El-Tahlawy 1999; Zhang et al. 2003). But, their applications in textiles have not yet been well studied. The practical application of such nanochitosan to textiles at shop floor level demands suitable technology for the production of nanochitosan dispersions, characterization and the analysis of stability of standing baths. Therefore, attempt has been made to set a simple methodology to produce nanochitosan by ionotropic gelation with pentasodium tripolyphosphate. Chitosans of different molecular weights are obtained by controlled depolymerization of parent chitosan using nitrous acid hydrolysis, and these products are subsequently used for the synthesis of nanochitosan. The final results of many basic experiments after getting the confirmation of reproducibility are given herein. A representative concentration of 1gpL is, therefore, only reported and discussed here to make it as brief as possible and to avoid presenting less important data. The influence of particle size on different properties of nanochitosan-treated cotton fabric, like appearance, stiffness, absorbency, dyeing behaviour, and wrinkle recovery, has been explained. SEM images of nanochitosan-treated cotton fabric have also been analysed.

7.3 Technical Details

The following materials have been used

- 100 % cotton fabric ready for dyeing stage
- Certain class of direct dyes
- Chitosan (CHT1), having degree of deacetylation (DAC) 90 % and viscosity 22 cPs
- Dimethylol dihydroxy ethylene urea (DMDHEU) and other chemicals such as sodium tripolyphosphate (TPP), acetic acid, sodium nitrite, sodium acetate (anhydrous), and sodium hydroxide used were of analytical grade.

Chitosans with various molecular weight grades have been first obtained by depolymerization of CHT1 by nitrous acid hydrolysis method, which have been employed for the preparation of nanochitosan dispersions as described elsewhere (Tiwari and Gharia 2003). In general, chitosan was dissolved in acetic acid solution, and an optimized quantity of TPP was added drop-wise with rapid stirring (about 400 rpm) to obtain an opalescent solution. The sample was allowed to stand overnight, filtered through sintered glass filter of porosity grade G3, and preserved in refrigerator. The prepared nanochitosan was termed as CHT1 N (Chattopadhyay and Inamdar 2013). The synthesized nanochitosan was applied to cotton fabric within 24 h since the stability of nanochitosan gets adversely affected with time as discussed earlier. The various grades of chitosan and nanochitosan have been specified. A particle size analyser has been used to determine the particle size and size distribution of the chitosan.

A two-dip–two-nip technique has been used to apply nanochitosan dispersion (1 gpL) onto fabric, using a padding mangle with wet pickup of 70 %. After drying, the fabric was cured in oven at 150 °C for 4 min. The sample was then washed in the following sequence: rinse → alkali wash (soda ash 1 gpL, MLR 1:50) → hot wash (twice) (85 °C/20 min) → cold wash → dry.

The following studies have been done

- SEM
- Evaluation of Indices
- Fabric Stiffness, Tenacity and Absorbency
- Crease Recovery Angle and Antimicrobial Activity.

7.4 Synthesis and Characterization of Nanochitosan

Chitosan possesses quite a long linear structure and has rigid conformation. The CHTI (Parent chitosan sample 1) has a characteristic size of hydrodynamic sphere of 4014 nm. The higher viscosity of the solution is due to the higher particle size in the solution. Any molecular size chitosan can reduce the particle size to nanolevel by 'bottom-up' approach (Patel and Jivani 2009). Due to its polycationic nature, chitosan is subjected to ionic gelation with poly anions like pentasodium tripolyphosphate (TPP), ethylene diamine tetra acetic acid (EDTA), resulting in formation of nanoparticles. Such particles are stabilized by electrostatic hindrance due to coulombic repulsion between particles of same ionic charges (Zhang et al. 2010; Trapani et al. 2009; Chattopadhyay and Inamdar 2010). Owing to faster ionic reactions between chitosan and TPP, non-toxic nature of these components and ease of operation, we adopted the gel ionization technique for the synthesis of nanochitosan particles. The particle size distribution of CHT5 N, having particle size of 110.74 nm, is given in Fig. 15. Scaling down the particle size of large polymeric materials to nanolevel is a big challenge. It is clear from them present study that the molecular weight (Table 1) has a great role in controlling the particle size and by reducing the molecular weight we achieved about 110-nm particle size. The study evidently shows that the particle size can be decreased below 100 nm by conducting a trial with parent chitosan of low molecular weight (lesser than 10,000). This investigation would provide a platform for future work of such a type and would serve basic information to the future researchers.

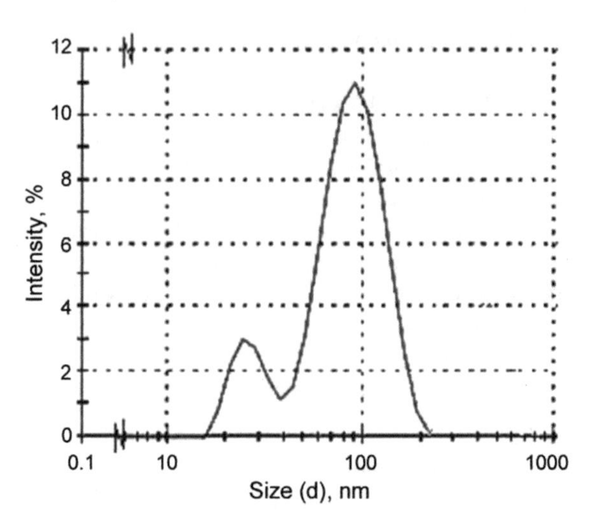

Fig. 15 Nanochitosan size distribution by intensity (CHT5N) (Chattopadhyay and Inamdar 2013)

7.5 Influence of Nanochitosan Application on Surface Morphology

Treated as well as untreated cotton has been investigated for surface morphology using SEM microscope (Fig. 16). Chitosan exhibits an inherent property of film formation, which is clearly seen as gloss on fibre surface as shown in Fig. 16(b) (Chattopadhyay and Inamdar 2013). Further, the film deposition on fibre surface can be confirmed by prolong boiling of treated sample in distilled water so that the broken appearance of film can be viewed under SEM (Fig. 16c). Fabrics treated with nanochitosan exhibit a totally different microphotograph (Fig. 16d–f). Nanochitosan film has more uniformity.

7.6 Influence of Nanochitosan on Fabric Appearance

The appearance and the feel of the fabric dictate its appeal. The influence of particle size of nanochitosan on these properties of cotton fabric has been studied. The appearance and the fabric feel are quite satisfactory. The whiteness is improved with reduction in particle size and reaches well nearer to that of control sample (Chattopadhyay and Inamdar 2013). This may be attributed to the greater extent of penetration of nanochitosan particles into fibre structure and allowing the cuticle for exposure. Normal chitosan forms a film on surface, and thereby changes the whiteness to a certain degree. This film may also impart stiffness to the fibre, whereas a nanochitosan shows a little influence, as per the studies done.

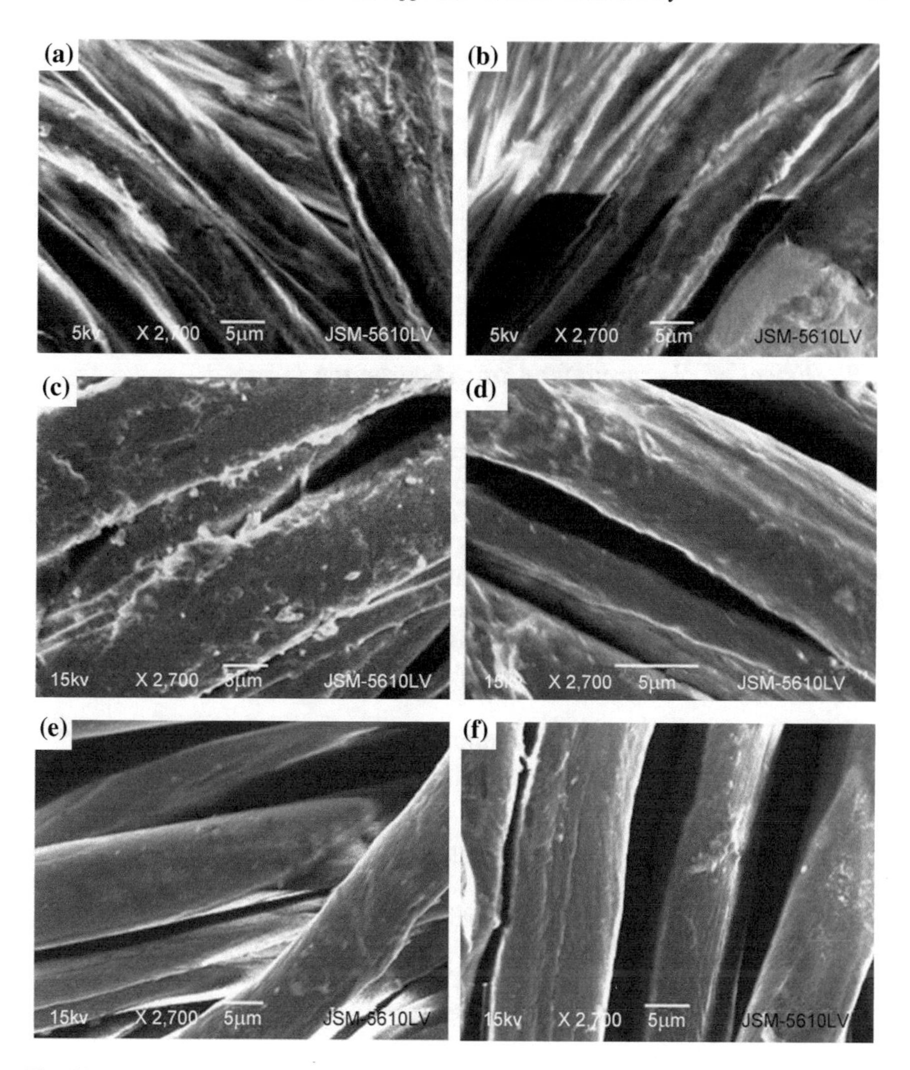

Fig. 16 SEM images (×2700) of **a** cotton fibre (control), **b** CHT1-treated fibres, **c** CHT1-treated and then prolong boiled cotton fibres, **d** CHT1N(ii) 319.4 nm, treated cotton fibres, **e** CHT4N, 195.2 nm, treated cotton fibres and **f** CHT5N, 110.74 nm, treated cotton fibres (Chattopadhyay and Inamdar 2013)

7.7 Influence of Nanochitosan on Tensile Properties

The influence of nanochitosan treatment on tensile properties of cotton fabrics has been investigated (Chattopadhyay and Inamdar 2013). A decrease in the strength takes place owing to application of conventional chitosan. Conventional chitosan mostly forms a film on the surface of the fabric and very less amount of it can enter into the interfibre regions and thus cannot take part in load bearing phenomenon, rather affects symmetrical distribution of load. Nanochitosan, on the other hand, because of its small size can easily enter into the interfibre region and to even inter-cellulosic chain regions and work as a cross-link, which bears the load to a great extent. The strength improvement is therefore clearly observed with the reduction in particle size. But the elongation property gets reduced to a certain degree due to the scaling down of particle size. Due to the in situ formation of 3D networks the adjacent fibre molecules show possible resistance to slip and reduce the elongation-at-break.

7.8 Effect of Nanochitosan on Absorbency

Figure 17 depicts the absorbency determined by drop penetration method pertaining to nanochitosan-treated cotton fabric. The results show that the absorbency is decreased with the reduction in particle size. This may be elucidated by the example of lotus leaf effect. Distribution of nanochitosan particles as a thin layer over and beneath the surface, (Fig. 16d–f), may roll out the water droplets. The absorbency of nanochitosan-treated fabrics continues to fall within the tolerable limits of conventional wet processing conditions (Chattopadhyay and Inamdar 2013).

7.9 Dyeing Behaviour of Nanochitosan-Treated Cotton Fabric

As the chitosan structure is very much similar to cellulose, its treatment to cotton is expected to influence the dyeing. Thus, the influence of pretreatment of nanochitosan on direct dyeing of cotton has been investigated. The influences of chitosan and nanochitosan pretreatment on dye uptake have been determined. Due to the decrease in the particle of chitosan, the dye uptake by treated cotton fabric, in conventional process, is increased progressively. The results are superior to

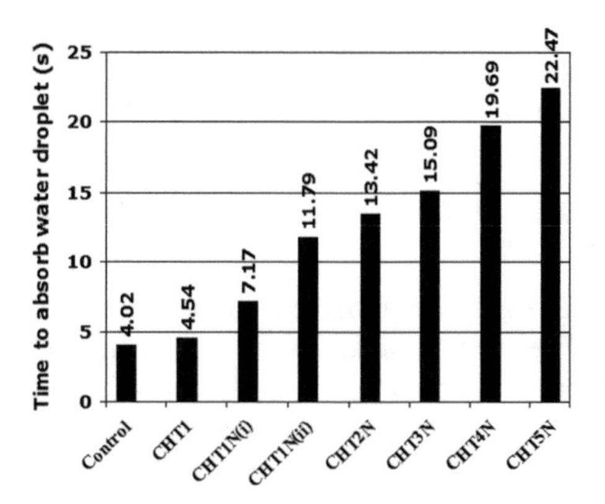

Fig. 17 Influence of particle size on treated cotton fabric absorbency (Chattopadhyay and Inamdar 2013)

corresponding parent chitosan-treated materials. The dye uptake of CHTN-treated samples is found to be significantly increased, resulting in almost complete exhaustion of dye bath, when acidification was followed (Chattopadhyay and Inamdar 2013). The increased dye uptake due to chitosan treatment may be attributed to the presence of primary amino groups of chitosan. These cations dissipate the negative surface charge on cotton and drives dye molecules to the fibre. Further, the dye uptake may also been enhanced due to the dyeability of chitosan itself with direct dyes. The nanochitosan due to increased surface area and hence higher accessibility for dye sites put much added value. The primary amino groups on chitosan get protonated (quaternized) in acidic medium having enhanced positive charge, thus forming salt linkages with anionic (sulphonate) groups of residual dye in the bath. Secondly, the higher dye uptake value after acidification proves the presence of chitosan. The wash fastness and rub fastness properties of direct dyed fabrics were also analysed. As the particle size is reduced, the fastness to washing is improved, which could be considered as the formation of chitosan–dye complex in situ. The rubbing fastness is also improved to a certain degree with the particle size decrease.

7.10 Influence of Nanochitosan on Crease Recovery

The aesthetic appeal of garment/fabric gets downgraded due to susceptibility to creasing. In the conventional method, the problem is overcome by treatment of cotton fabric with different cross-linking agents based on aminoplast resins, e.g. DMDHEU (Lopez-Leon et al. 2005; Boonyo et al. 2008; Loretz and Bernkop 2006). The crease recovery property as a function of chitosan and nanochitosan treatments has been compared against DMDHEU. The crease recovery angle of cotton fabric is greatly reduced by the treatment of normal chitosan (CHT1). Treatment of cotton fabric with chitosan of lower particle size is found to improve the crease recovery of cotton fabric. However, yet it could not gain the rating of commercially used cross-linking agent DMDHEU. Conventional chitosan is believed to form a surface coating which lowers the possibility of cross-linking and therefore cannot contribute to the load-sharing phenomenon. Owing to the greater penetration of the nanochitosan into the fabric structure, the improved wrinkle recovery property is improved. These polycationic nanoparticles may be bound to the fibre molecules and resist creasing to a certain degree, owing to their better penetration.

7.11 Influence of Nanochitosan on Antibacterial Effect

Owing to their moisture and warmth, cotton fibres are similar to other natural fibres in creating a conducive environment for the growth of microorganisms. The organisms cause discolouration, development of rancid/bad odour, stains and strength loss of fabric, skin allergies, and also infectious diseases in human body (Knittel and Schollmeyer 2002; Lopez-Leon et al. 2005). One of the most popular ways of imparting antimicrobial resistance is to use nanosilver colloid. Chitosan, being polycationic material, binds to anionic surfaces of microbe cell wall and disrupt it leading to death of cell (Loretz and Bernkop 2006). Attributing to the antibacterial and metal particle retention properties of chitosan (Du et al. 2009), the fabric was treated with chitosan and nanochitosan and then with nanosilver colloid. Nanosilver colloid with 1×10^{-3} M/100 mL concentration and average particle size 110 nm has been prepared as reported elsewhere (Loretz and Bernkop 2006). The resistance against bacterial attack of untreated and treated samples of cotton was determined by measuring the loss in strength due to soil burial test. The chitosan can be used as an efficient antibacterial agent. There is improved effect due to the decrease in particle size of nanochitosan and coupling with nanosilver.

8 Printing of Polyester Fabric with Disperse Dye Nanoparticles

8.1 Overview

Disperse dye has been ground to a nanosize of 23 nm and then exposed to ultrasound treatment, in the form of a solution with two dye concentrations for 8 h. The nanoparticle suspension is directly added to the stock thickener and printed on a polyester substrate. All measurements as well as various parameters affecting dye fixation are investigated. The results indicate that grinding and ultrasound exposure of the dye results in minimizing particle size, which facilitates dye penetration in the hydrophobic substrate (Osman and Khairy 2013). This result is obtained without the incorporation of a carrier in the printing paste and only pH is adjusted at 6. Very good fastness properties are seen in the dye fixation of the prints by steaming at 130 °C for 30 min, and no differences are seen from prints of untreated dye with ultrasound. The influence of grinding on size of dye particle is shown in SEM measurement.

8.2 Related Aspects

Disperse dyes are those which are water insoluble have affinity towards certain hydrophobic fibres (Schindler and Hauser 2004). They conform to the non-ionic class of dyes, and are the generally used dyes in the textile industry for dyeing synthetic fibres like polyester, acrylic, and acetate (Gao and Cranston 2008; Du et al. 2009). The water solubility of disperse dyes is very low and hence should be milled to a very low particle size and dispersed in water using a surfactant (dispersing agent) or a carrier must be added during textile colouration (Chattopadhyay and Patel 2009). The actual mechanism by which a carrier used in dyeing accelerates textile colouration has been widely debated. Polyester fibres absorb the carrier and swell. This swelling can impede liquor flow in packaging causing unlevelness. The overall effect leads to lowering of the polymer glass transition temperature (T_g), thus promoting polymer chain movements and creating free volume. This speeds up diffusion of the dye into the fibres. Alternatively, the carrier may form a liquid film around the surface of the fibre in which the dye is very soluble, thus increasing the rate of transfer into the fibre (Harrocks and Anand 2000). Power ultrasound (US) can enhance a wide variety of chemical and physical processes, mainly due to the phenomenon known as cavitations in a liquid medium that is the growth and explosive collapse of microscopic bubbles. Sudden and explosive collapse of these bubbles can generate hot spots (Lachapelle and Maibach 2009), i.e. localized high temperature, high pressure, shock waves, and severe shear forces capable of breaking chemical bonds. Many efforts have been made explore this technique in the textile colouration as it is a major wet process, which consumes much energy and water and releases large effluent to the environment. The phenomenon of cavitations lead to improvements in ultrasound-assisted colouration processes, and hence other mechanical and chemical effects are as follows:

- Dispersion (breaking up of aggregates with high relative molecular mass);
- Degassing (explosion of dissolved or entrapped air from fibre capillaries);
- Diffusion (accelerating the rate of diffusion of dye inside the fibre);
- Intense agitation of the liquid;
- Destruction of the diffusion layer at dye/fibre interfaces;
- Generation of free radicals; and
- Dilation of polymeric amorphous regions.

8.3 Technical Details

Disperse dye with nanosized particles has been printed onto 100 % polyester fabric. The dye has been milled and treated with ultrasound for various time durations so as to get tiny dye particles dispersible in the printing paste without adding a dispersing agent, as the size of nanoparticles is small enough to diffuse into the hydrophobic fibres properly. All materials used were of analytical grade (Osman and Khairy 2013). The disperse dye has been ground into finer particle size. The dye powder was sealed in a hardened steel vial (AISI 44 °C stainless steel) using hardened steel balls of 6 mm diameter. Milling was performed using a ball–powder mass ratio of 4:1. The dye was milled at different intervals, such as 4, 6, 9 and 25 days. After each milling interval, the particle size of the resulted dye powder was measured. The smallest particle size of 23 nm chosen was obtained from milling the dye powder for 25 days. Two stock solutions were prepared using the milled nanoparticle dye powder of 1 and 3 %, where 1 and 3 g of dye powder was dispersed in 99 and 97 mL of distilled water respectively. Each dispersion has been irradiated with ultrasound waves (720 kHz) and stirred at 80 °C for various time durations.

To investigate each factor, two printing pastes were prepared for each parameter containing the two different dye concentrations. The pH was adjusted at 6 using sodium dihydrogen phosphate. The printing paste was applied to fabric through flat-screen printing technique and then samples were left to dry at room temperature. Fixation of dye was carried out with two methods, namely steaming at 130 °C for 30 min and thermofixation at 140 °C for 10 min to determine the optimal fixation method that results in best K/S values. The fabrics were finally washed off using a 2 g/L non-ionic detergent (Sera Wash M-RK) at a liquor ratio of 1:50. Soaping treatment has carried out at 60 °C for 10 min.

8.4 Influences of Ultrasonic Irradiation and Particle Size

Solutions with two different nanoparticle dye concentration have been exposed to ultrasound irradiation for 4, 6 and 8 h, respectively. Figure 18 depicts the colour strength values of the prints, both steamed and thermofixed. It is obvious that ultrasound treatment of both stock dye solutions has a great influence on the colour yield of the printed polyester fabrics. To determine the best dye fixation method, steaming and thermofixation are applied on the prints separately using both dye concentrations. For steamed prints, optimum K/S enhancement reaches 85.4 and 53.9 % for polyester prints with 1 and 3 % stock solutions respectively compared with the untreated dye solution with ultrasound waves. These great results are due to the fact that ultrasound enhances the dye molecule's diffusion into the fibres

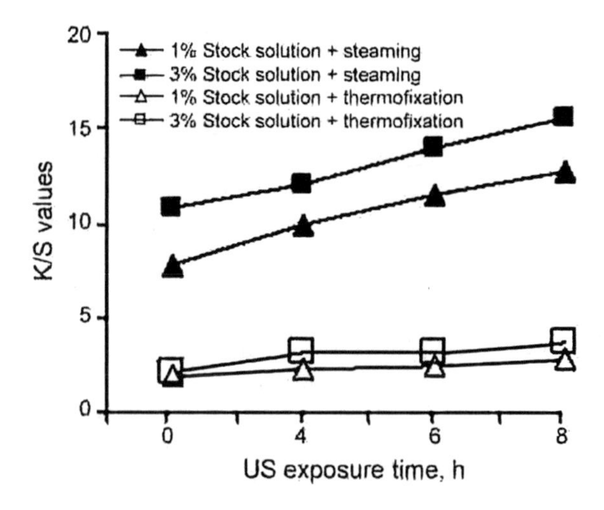

Fig. 18 Influence of ultrasound exposure duration on *K/S* values of steamed and thermofixed polyester prints (Osman and Khairy 2013)

(Broadbent 2001; Fung and Hardcastle 2001; Nahed and El-Shishtawy 2010). Also, ultrasound has a significant effect on the reduction of particle size of the disperse dye (Saligram et al. 1993; Ahmed and Loman 1996). On the other hand, the figure shows that thermofixation has a negative influence on the *K*/*S* values of the prints. These results show that thermofixation in the absence of a carrier fixes only 50–70 % dye, while in the steaming process steam condenses on the cold fabric, raising its temperature to 100 °C and swelling the thickener film. The condensed water is largely evaporated again during the exposure to steam, but the thickener is not bound into the fabric as in dry fixation, subsequently handle of the fabric becomes softer. The absorption capacity (build-up) increases in proportion to the steam pressure and corresponding temperature. The depth of colour achieved at high pressures is not possible by using longer durations at lower pressures. The steaming is the best technique for dye fixation and has thus been selected to be applied.

8.5 pH of Printing Paste

As disperse dyes are sensitive to alkalis, polyester is normally dyed under acidic condition (Ahmed and Loman 1996). Sodium dihydrogen phosphate is recommended because, unlike some organic acids, it has no corrosive effect on nickel screens and is compatible with natural thickeners. Figure 19 illustrates the effect of printing paste pH using the nanotreated dye with ultrasound exposure for 8 h on the *K*/*S* of polyester fabrics. The printing process is carried out using two stock dye solutions having 1 and 3 % disperse dye concentrations printed on polyester fabric and followed by steaming at 130 °C for 30 min. The best values of colour strength are achieved in the case of both stock dye solutions at pH 6, which implies that the use of a nanoparticle disperse dye helps in dye penetration into the fibres more easily at more neutral pH value.

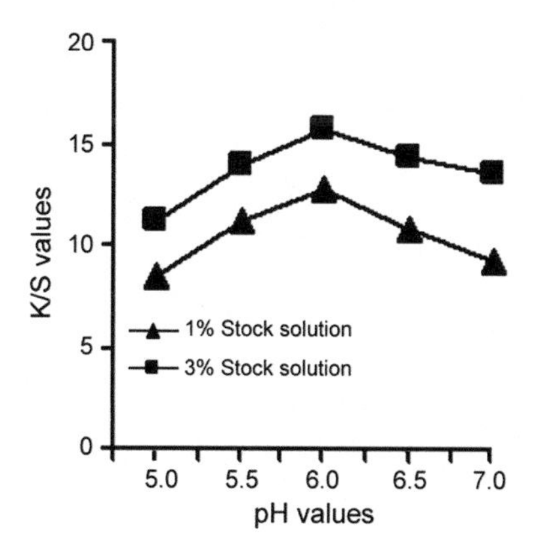

Fig. 19 Influence of pH of printing paste on the colour strength of polyester prints (Osman and Khairy 2013)

8.6 Steaming Conditions

The fixation is incomplete even though using saturated steam at 100 °C, if carrier is not used. In addition, the colour depth achieved at high pressures is not possible with longer durations at lower pressures (Lee and Kim 2001). Fibres of the most common polyester are quite crystalline, very hydrophobic, and have no ionic groups. Hot water does not swell them, and large dye molecules (that have lower diffusion coefficients) do not easily penetrate into the fibre interior.

For the samples used in this study, the effect of steaming temperature and time on the *K/S* of polyester fabrics printed with disperse dye nanoparticles of the two different concentrations is investigated and the data is plotted in Fig. 20.

It is observed that the best *K/S* values are obtained by increasing both steaming temperature and time that reach their maximum values on steaming at 130 °C for 40 min. As a minor colour difference is seen during steaming for longer times, steaming for 30 min is selected as the best steaming time.

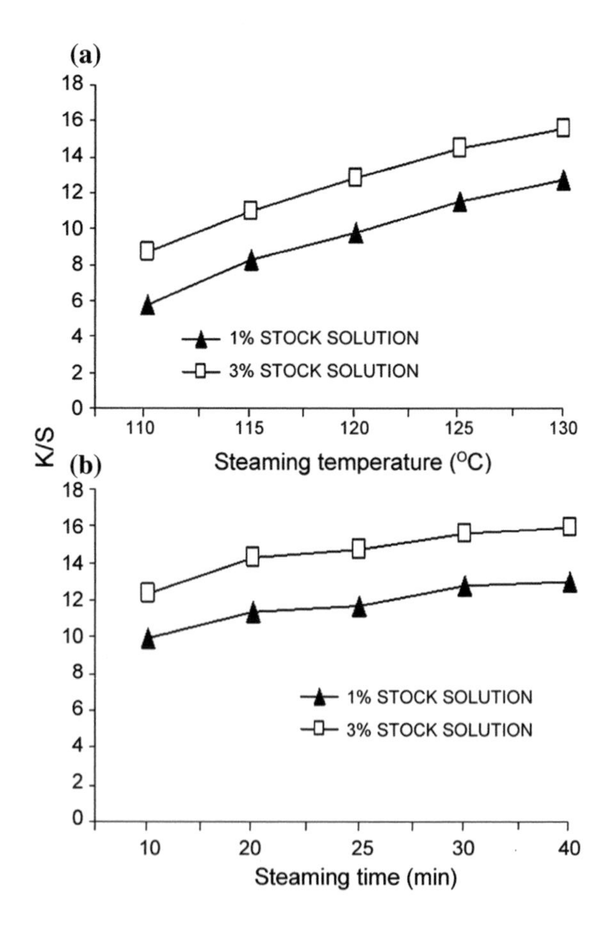

Fig. 20 Influence of steaming temperature (**a**) and time (**b**) on *K/S* of polyester prints (Osman and Khairy 2013)

Satisfactory results are obtained, as no carrier is added to the printing paste and are referred to both the nanosize of dye particles as well the as ultrasound treatment of dye.

8.7 *Investigations on SEM and TEM*

Figures 21 and 22 depict the surface morphology, structure, and particle size of dye samples milled at various time durations before being exposed to ultrasound. The SEM images of dye particles appear in different shapes like breaking dishes shape, spherical shape, and tiny sprinkled dots (Fig. 21). The TEM micrographs of the samples milled at 9 and 25 days compared with the unmilled sample are shown in Fig. 22. The micrographs indicate uniform spherical dye nanoparticles. The micrographs also indicate that an average size of the nanosized dispersed dyes is 200–23 nm in diameter. It is found that the particle size decreases as the period of grinding increases. The difference in particle size after grinding arises from dissociation owing to the impact of shear forces that act on dye particles in the ball mill which eventually converts the particle size from 200 nm (before milling) to 60 nm (after 9 days of milling) and 23 nm (after 25 days of milling).

9 Application of Nanopolysiloxane on Jute-Blended Fabrics

9.1 *Overview*

Nanopolysiloxane finishing on jute-blended fabric is found to have a positive influence on improvement in handle, as it forms uniform polymer film on jute fibre surface, which reduces surface friction and increases softness (Lakshmanan et al. 2014). The decrease in flexural rigidity of the fabric is better in nanopolysiloxane/nano + macropolysiloxane combination finishing compared with other combination finishings. Nano + micropolysiloxane combination finishing shows better crease recovery than other finishing combinations on jute-blended fabric. The bending rigidity decreases more in weft direction with nanopolysiloxane finishing and hence improves the inner softness of jute fibre better than that of cotton fibre. Finally, the combination of nano + micropolysiloxane finishing shows the prospect to improve the handle properties of jute-blended fabric.

9.2 *Related Aspects*

Jute is an important cellulosic fibre constituted mainly of alpha cellulose, hemicellulose, and lignin and finds application in packaging and sackings for the food

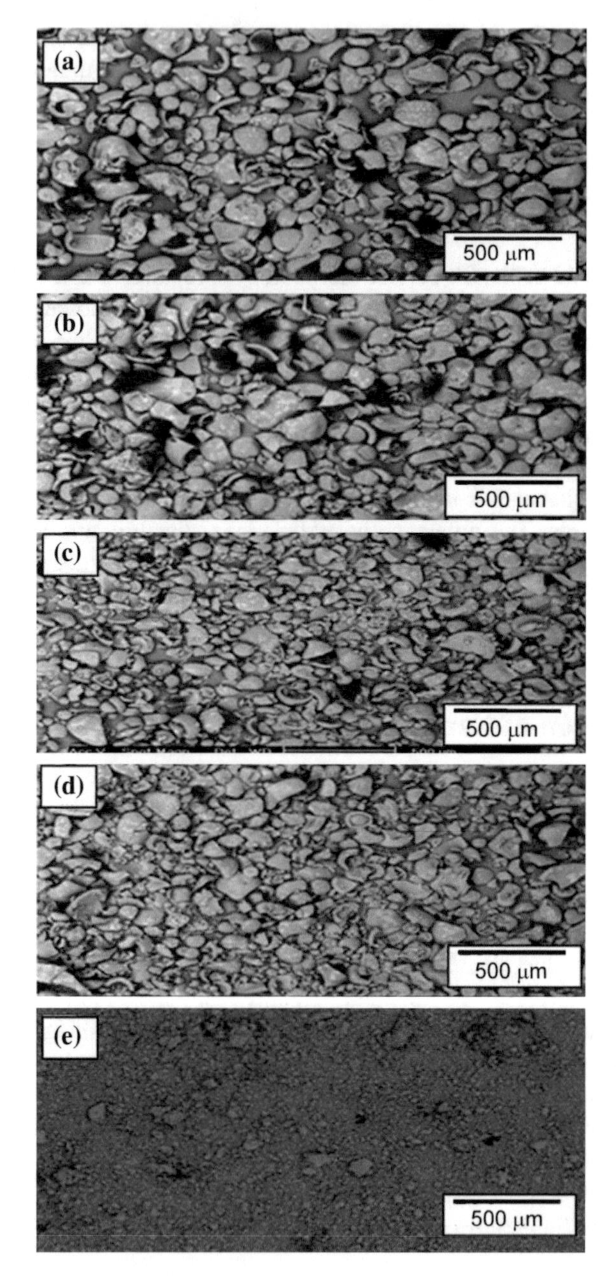

Fig. 21 SEM images of disperse dye **a** before milling, and after **b** 4 days, **c** 6 days, **d** 9 days, and **e** 25 days of milling (Osman and Khairy 2013)

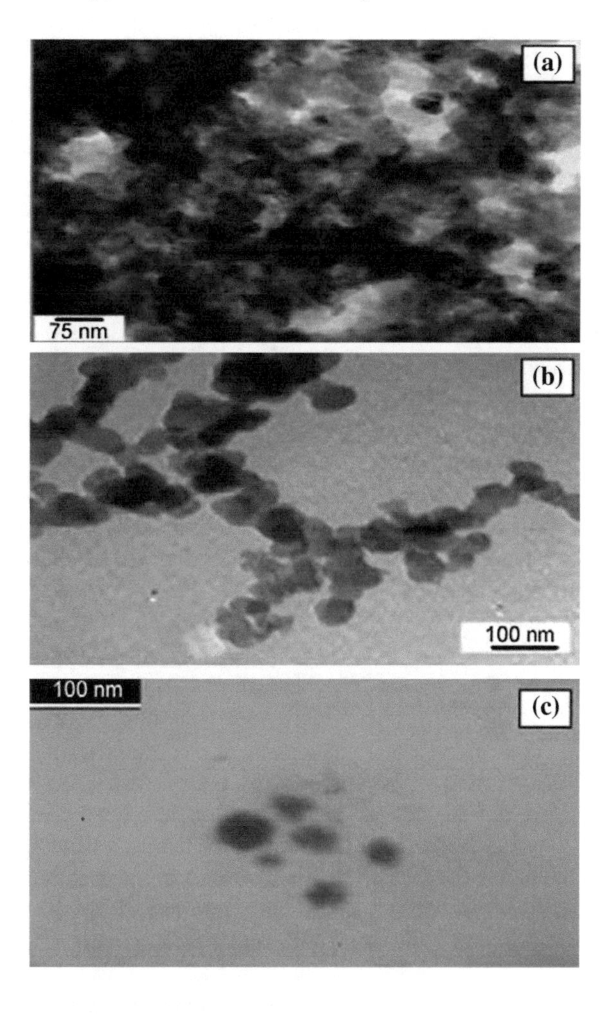

Fig. 22 Representative TEM images of disperse dye **a** before milling, and after **b** 9 days, and **c** 25 days of milling (Osman and Khairy 2013)

grains. But its unique properties such as roughness, coarseness, and stiffness arise due to its multicellular fibre structure which is bonded with hemi-cellulose and lignin (Leslie 1994; Kundu 1956). These properties generally create problem during fabric manufacturing as well as interfering in the performance of a final product. Due to this drawback, jute is not preferred in the apparel textiles. Development of fine yarn from jute fibre by blending with synthetic fibres, cotton or staple viscose rayon is one of the possible ways to prepare jute-blended apparel. A fine jute/polyester (70/30) blended yarn with linear density of 122 tex was prepared on a conventional jute spinning system, which was optimized by varying linear density, twist and blend proportion of jute and hollow polyester fibre (Rowell and Stout 1998; Debnath et al. 2007a, b). It was woven with cotton yarn as warp in a handloom for the development of union fabric. The physicomechanical properties of the fabric were evaluated and compared with conventional apparel fabrics. The results inferred that the properties are matched with basic requirement of a winter garment and hence later winter garment from this jute-blended fabric as outer cover has developed (Debnath and Sengupta 2009). However, due to surface roughness of jute fibre, the handle of the fabric did not meet the required soft feel. Textile processes such as scouring, dyeing and finishing are normally focused on the value addition and performance enhancement of textile products (Debnath et al. 2011).

Chemical finishing that is able to make uniform film on the surface of jute fibre has potential to enhance the handle property of jute-blended fabric (Ammayappan et al. 2013). Polysiloxane-based chemical finishing can be applied to improve the surface softness and handle properties due to improvement in flexibility of fibre polymer. The softness of wool/cotton union fabric can be enhanced through application of polysiloxane of various sizes either individually as nano-, micro-, and macroemulsion or in combination of these forms (Ammayappan and Moses 2010).

Very little information is available on application of nanopolysiloxane finishing either in individual form or in combination form on jute-blended fabric. Hence efforts have been taken to apply nanopolysiloxane on jute-blended fabrics in 5 chemical finishing formulations adopting pad-dry-cure technique and comparing its performance.

9.3 Technical Details

Union fabric of jute–polyester-blended yarn (70:30 ratio) in weft direction and 100 s cotton yarn in warp direction has been used. To reduce the water consumption in the finishing, finishing chemicals were applied by dry-on-wet method, and hence the fabric was used without any pretreatment at pH 5 (Lakshmanan et al. 2014). After curing, fabric was conditioned, rinsed with distilled water gently to remove non-ionic detergent, and dried at ambient condition. The properties of finished and unfinished samples such as finish add-on, bending length, flexural rigidity, and dry crease recovery angle were evaluated as per the standard procedure (British Standards 2012; American Association of Textile Chemists and Colorists 2003). SEM has been used to magnify control and finished jute fibre samples for surface investigation.

9.4 Finish Add-On

The amount of finishing chemical added on the jute-blended fabric is shown in Table 2. The percentage finish add-on is higher in nanopolysiloxane-based finishing (3.56 %) than other nanopolysiloxane combination finishing (<3.08 %). The cell wall of swollen natural fibres consist of several hundreds of lamellae and have pores with a most common pore size of 160–380 nm (Parthasarathi 2008), while the

Table 2 Finishing combinations for jute-blended fabric (Lakshmanan et al. 2014)

Combination No.	Finishing chemical	Amount finish add		
		g/L	on %	Used
1	Ceraperm TOWI	80	3.56	Nano
2	Ceraperm TOWI + Ceraperm MW	40 + 40	3.08	Nano + Micro
3	Ceraperm TOWI + Ceraperm UP	40 + 40	2.98	Nano + Macro
4	Ceraperm TOWI + Leomin HBN	40 + 40	3.05	Nano + CS

size of nano-, micro-, and macropolysiloxane emulsion ranges in 50–100, 200–300, and >500 nm, respectively (Lakshmanan et al. 2014).

During application, most of the nanopolysiloxane emulsion can be easily adsorbed on the surface of the fibre and then diffused inside the fibre matrix; micropolysiloxane emulsion can be penetrated inside the fibre partially; macropolysiloxane mainly can be spread only on the surface of the fibre, while cationic softener can be coated on the surface of the fibre (Ammayappan and Moses 2010). Due to improved diffusion behaviour of nanopolysiloxane, the extent of deposition is higher than its combination finishing. Moreover, in combination finishing, the diffusion of nanopolysiloxane to inside the fibre is hindered, leading to a reduction in finish add-on.

9.5 *Dry Crease Recovery Angle*

Figure 23 depicts the dry crease recovery angle of finished and unfinished jute-blended fabric in both directions. Jute and cotton fibres have many free –OH groups in their fibre matrix. When treated with prepolymer of nanopolysiloxane

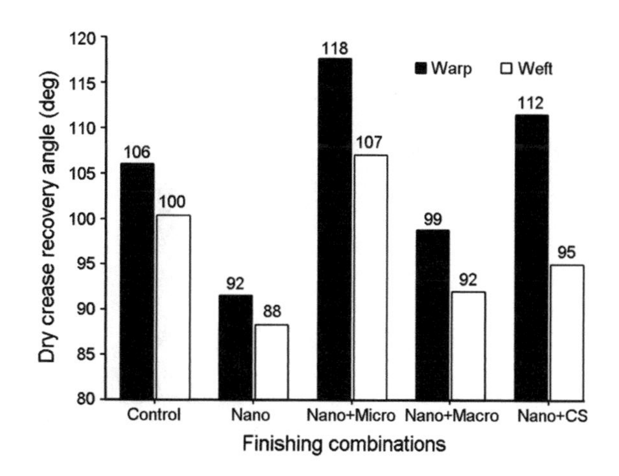

Fig. 23 Dry crease recovery angle of jute-blended fabrics finished with nanopolysiloxane-based finishing (Lakshmanan et al. 2014)

emulsion, they are polymerized in the form of a thin film on the surface of the fibre, thus masking the free –OH groups. The improvement in crease recovery angle is observed in nano + micro- and nano + CS combination finished fabrics when compared with untreated fabric in both directions (British Standards 2012; American Association of Textile Chemists and Colorists 2003; Steele 1962; Bajaj 2002). The crease recovery property of the fabric in improved both in warp (10 %) and in weft (7 %) directions by combination of micropolysiloxane finishing.

Since nanopolysiloxane improves the inner softness of jute and cotton fibres, it does not show positive improvement in crease recovery. But, owing to the presence of jute fibre in weft yarn, the dry crease recovery angles of both finished and unfinished fabrics in warp direction are higher than weft direction.

9.6 Bending Length

Figure 24 depicts the bending length of finished and unfinished jute-blended fabrics in both directions. The bending length of control fabric is greater in weft direction than in warp direction because of linear density variation between warp and weft yarn and also stiffness of jute fibre in weft direction. In warp direction, nano + macropolysiloxane emulsion combination finishing shows more reduction in bending length (10 %) than that of nanopolysiloxane-based finishing (5 %), while in weft direction, nanopolysiloxane finishing shows more reduction (12 %) than that of other finishing combinations (4–9 %). Nanopolysiloxane diffuses well inside the fibre matrix and forms a polymer networking, which improves the softness of fibrils of fibres, i.e. inner softness of jute fibre, and hence nanopolysiloxane finishing reduces the bending stiffness of the jute fibre (Saville 1999). But in the case of combination finishing, the improvement in inner softness by the nanopolysiloxane may be reduced, and hence the reduction in bending length is lesser in warp direction compared to weft direction. The nanopolysiloxane is

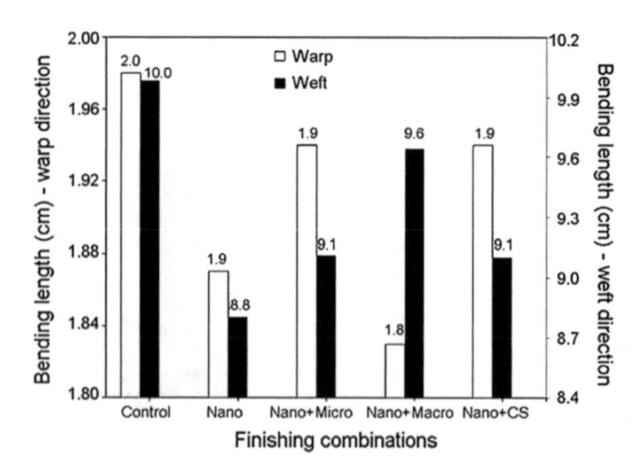

Fig. 24 Bending length of jute-blended fabrics finished with nanopolysiloxane-based finishing (Lakshmanan et al. 2014)

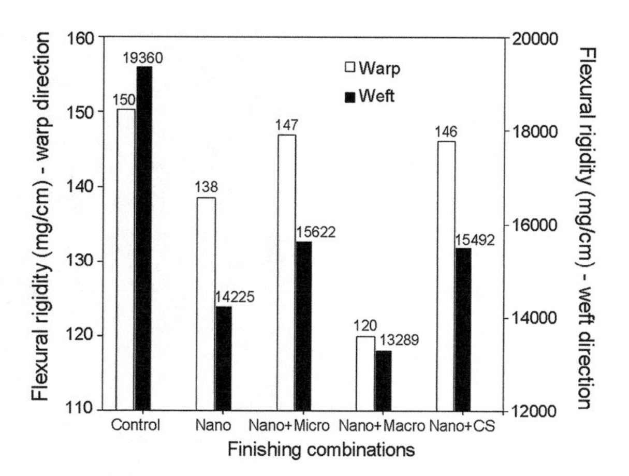

Fig. 25 Flexural rigidity of jute-blended fabrics finished with nanopolysiloxane-based finishing combinations (Lakshmanan et al. 2014)

found to have better effect in the reduction in bending stiffness in case of weft yarn (jute–polyester yarn) than in case of warp (cotton yarn) (Fig. 24).

9.7 Flexural Rigidity

Figure 25 depicts the flexural rigidity of finished and unfinished jute-blended fabrics in warp and weft directions.

Flexural rigidity is a measure of resistance of a cloth against bending by external forces, and it correlates with the weight per unit area and bending length of the fabric (Ammayappan et al. 2011; Ren et al. 2008). Being less-elongated natural fibre, jute fibre has more stiffness than cotton fibre, and hence the flexural rigidity of blended fabric in weft direction (<19,360 mg/cm) is higher than in warp direction (<150 mg/cm) both in control and in finished fabrics. Compared with other finishing combinations, the reduction in the flexural rigidity is better in nano + macropolysiloxane combination finishing both in warp (20 %) and in weft directions (31 %).

9.8 Cost-Effectiveness

Dry-on-wet method has been used to apply the finishing chemicals on the fabric and the findings have been compared with wet-on-wet method (commercial). In dry-on-wet method, by using 100 litres of finishing liquor, it is estimated that nearly 143 kg of dry fabric could be finished by each finishing formulations with 70 % expression, while in wet-on-wet method by using 100 litres finishing liquor nearly 500 kg wet fabric can be finished, since it took 20 % expression of finishing liquor

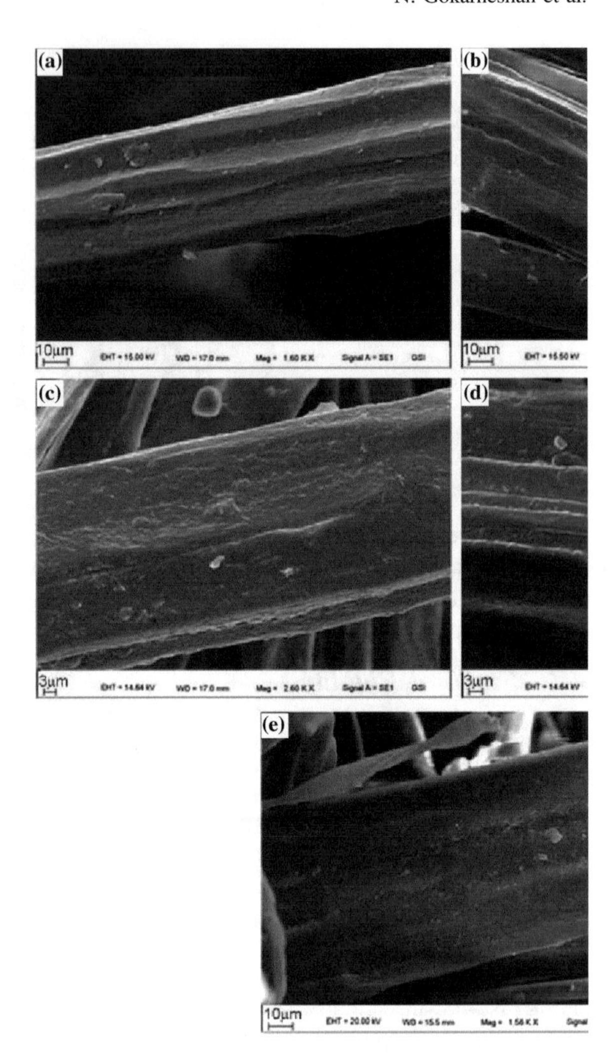

Fig. 26 SEM images of jute fibre finished with different finishing combinations **a** control, **b** nanofinished, **c** nano + microfinished, **d** nano + macrofinished, **e** CS + nanofinished (Lakshmanan et al. 2014)

(Lakshmanan et al. 2014). Considering the cost, the nano + macro- or nano + micropolysiloxane finishing combination in wet-on-wet condition on jute-based textiles could possibly decrease the cost of finishing with finishing performance that is durable.

9.9 SEM Investigation

Figure 26 shows the SEM images of control and other 4 nanopolysiloxane-based finished jute fibre. It is inferred that untreated jute have irregular grooves on the surface of the fibre, and after finishing, each finishing formulation forms a polymer film on the surface of the fibre. The coating and coverage of grooves are better in nanopolysiloxane finished fibre than in other finishing combinations (Lakshmanan et al. 2014). There is greater coverage than in other finishing combinations, as the add-on is higher in nanopolysiloxane finishing.

10 Nanometal Oxides in Viscose Fabric Modification

10.1 Overview

FTIR image of untreated viscose fabric as well as fabrics pretreated with 3-bromopropionic acid and then with nanometals tends to promote the reaction between the viscose fabrics and 3-bromopropionic acid. During the modification of viscose fabrics, the alkali-combining capacity values increase remarkably with the increase in the amount of 3-bromopropionic acid during the modification of the viscose fabrics, which supports the reaction between –OH active group of viscose and Br halide of 3-bromopropionic acid (El-Sayeed et al. 2015). The ability of the introduced carboxylic groups has been shown to attract the nanometal oxides from their suspension to the fabrics. The study also shows the incapability of the 3-bromopropionic acid-treated viscose fabrics to resist microorganism's growth. A unique ability to stop growth of these microorganisms on the viscose fabric can only be attained when treated with 3-bromopropionic acid followed by after treatment with nanometal oxides. It is also obvious that the ability of nanometal-treated viscose fabrics to reduce the microbial growth is in the following order: zinc oxide > aluminium oxide > titanium (IV) oxide. 3-Bromopropionic acid has a superior antifungal activity in comparison with its antibacterial activity. The added carboxylic groups to the fabrics acts as attracting groups for the nanometal oxides and also fixes these nanometal oxides to the fabric. A high antimicrobial activity has been observed even after 30 wash cycles in the viscose fabrics treated with 3-bromopropionic acid and nanometal.

10.2 Related Aspects

It is well known that textiles are a suitable substrate for the growth of microorganisms, particularly at appropriate humidity and temperature in contact with human body. The increasing public awareness on hygiene has prompted many investigations on antimicrobial agents for textiles. These agents are used to prevent serious undesirable effects on textile materials, such as degradation of colouring, staining and deterioration of the fibres, formation of unpleasant odour, increasing potential health risks (Dastjerdi et al. 2009, 2010; Hasebe et al. 2001; Bagherzadeh et al. 2007; Montazer and Afjeh 2007; Gao and Cranston 2008). A proposal of hygienic living standard by controlling the microorganisms is necessary. Researchers have focused on the antibacterial and antifungal modifications of textiles (Ladhari et al. 2007; Nakashima et al. 2001; Perelshtein et al. 2008; Shin et al. 1999; Yang et al. 2003; El-Sayed et al. 2010, 2012; Kantouch and El-Sayed 2008; Kantouch et al. 2013; Mekewi et al. 2012; Fu et al. 2005); by application of inorganic nanotechnology (Wong et al. 2006b; Daoud and Xin 2004; Daoud et al. 2005; Li et al. 2007; Tong et al. 2003; Ki et al. 2007; Liu et al. 2008; Parikh et al. 2005).

As most textile fabrics would undergo repeated laundering during their lifetime, the washing durability of nanometal-treated fabric is of significant importance. Polycarboxylic acids are multi-functional organic molecules with chemical and thermal stability (Bendak et al. 2008; Salama et al. 2011). Polycarboxylic acids could form ester linkage with hydroxyl groups of cellulosic fabrics at elevated temperature above 160 °C (Barari et al. 2009). They have also been used to improve the adhesion of the inorganic–organic interface (Huang et al. 2011).

The present work is aimed at preparing permanent antimicrobial viscose fabrics by fixation of propionic acid groups at lower temperature (below 100 °C), as active centres, onto the cellulosic polymeric chain. In the case of certain oxides like titanium oxide, zinc oxide, or aluminium oxide nanoparticles, the added carboxylic groups can possibly act as favourable centres. The antimicrobial efficiency with regard to the durable performance against selected microorganisms onto modified textile has also been evaluated.

10.3 Technical Details

Plain weave scoured 100 % viscose fabric has been used. The chemicals used have been of laboratory grade and include 3-bromopropionic acid (97 %), aluminium oxide nanopowder (particle size 50 nm), zinc oxide nanopowder (particle size 50 nm), and titanium (IV) oxide nanopowder (particle size 70 nm) (El-Sayeed et al. 2015). The fabric treatment involved two steps–pretreatment of viscose fabric with 3-bromopropionic acid, followed by treatment with metal nanopowder (aluminium oxide, zinc oxide or titanium (IV) oxide nanopowder suspensions). The following studies have been carried out

(a) FTIR
(b) TEM
(c) Antibacterial and antifungal activities
(d) Wash fastness
(e) Elution of metalized viscose samples—Amounts of the eluted metal ion were measured using atomic absorption spectrometer.

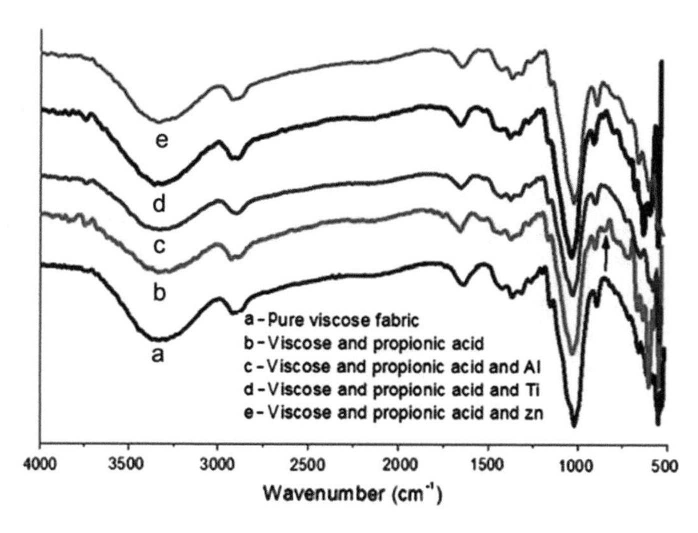

Fig. 27 FTIR image of viscose untreated fabric and fabric pretreated with 3-bromopropionic acid and nanometals oxide (El-Sayeed et al. 2015)

10.4 FTIR Investigations

Figure 27 depicts the FTIR image of untreated viscose fabrics and fabrics treated with 3-bromopropionic acid and nanometals. The spectral features of untreated viscose reveal that it is clearly distinguished by its broad distinguished peak of the hydroxyl groups at 3396.6 cm^{-1} (Fig. 27a). Formation of a new peak is observed, at 1560 cm^{-1} upon the treatment of viscose fabric with 3-bromopropionic acid (Fig. 27). This is a characteristic peak of the stretching vibration band of carboxylate groups introduced to the fabric surface (El-Sayeed et al. 2015). This new peak can be taken as support for the reaction mechanism between the viscose fabrics and 3-bromopropionic acid (Fig. 28). When compared with viscose treated with only 3-bromopropionic acid, there are no changes in spectral features of viscose fabrics pretreated with 3-bromopropionic acid followed by treatment with nanometal oxides (Ti, Al or Zn) (Fig. 27c–e).

10.5 Alkali-Combining Capacity of Viscose Fabric

The relation between the concentration of 3-bromopropionic acid used in pretreatment and the value of alkali-combining capacity (carboxyl content) of the modified viscose fabrics is depicted in Fig. 29. The findings show that the untreated viscose fabric has a moderate amount of carboxyl content (50/meq./100 g fabric), attributed

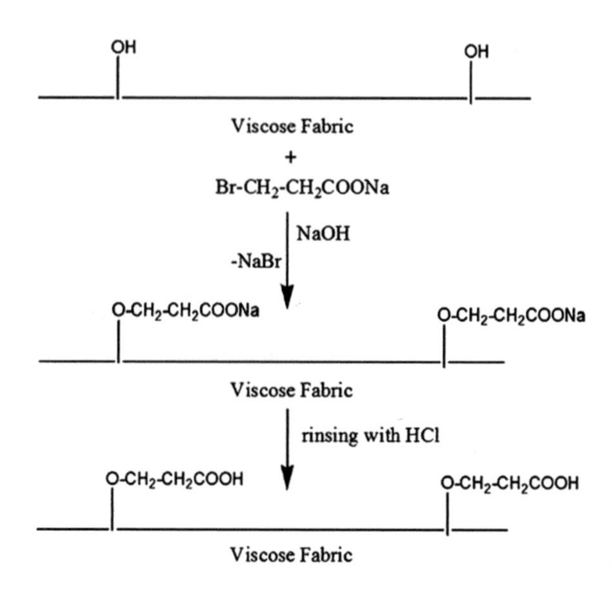

Fig. 28 Proposed reaction mechanism of 3-bromopropionic acid with viscose fabric (El-Sayeed et al. 2015)

to its nature and its manufacturing conditions. This value increases remarkably as the amount of 3-bromopropionic acid is increased during the modification. The progressive reaction between the hydroxyl groups of viscose and the bromine atom of the 3-bromopropionic acid has been supported by the obtained data.

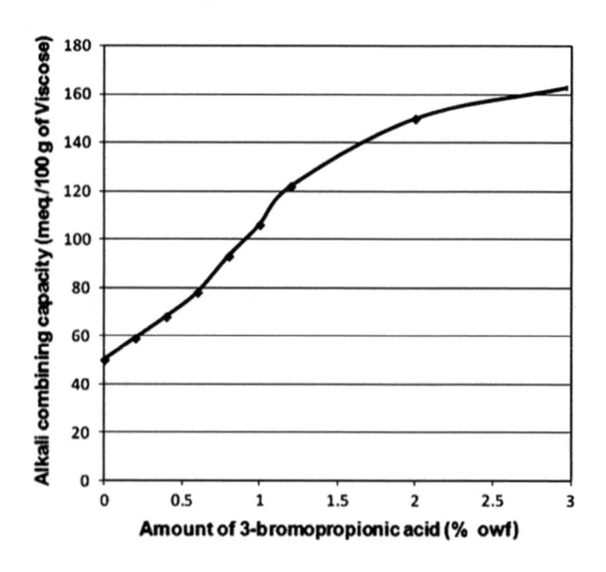

Fig. 29 Effect of concentration of 3-bromopropionic acid on alkali-combining capacity of pretreated viscose fabric (El-Sayeed et al. 2015)

10.6 Influence of 3-Bromopropionic Acid Concentration on Nanometal Oxides Uptake by Viscose Fabrics

It is well known that the carboxylic groups are able to attract and/or act as anchor on metal oxides, such as TiO_2 through electrostatic interaction (Campus et al. 1999; Dhananjeyan et al. 2001). The discussion herein shows the ability of the carboxylic groups introduced to the viscose fabrics to attract the nanometal oxides such as aluminium oxide, zinc oxide, or titanium (IV) oxide from their suspensions to the viscose fabrics. With the increase in the concentration of 3-bromopropionic acid, the amount of nanometal uptake by the viscose fabric is also increased (Fig. 30).

10.7 Antimicrobial Activity of Treated Viscose

Viscose fabrics untreated and treated with 3-bromopropionic acid, followed by treatment with nanometal oxides (aluminium oxide, zinc oxide, or titanium (IV) oxide), have been intended to destroy one of the most common Gram-negative bacteria (*E. coli*) as well as the famous fungus (*Candida albicans*). Figure 31a, b reveal the incapability of the 3-bromopropionic acid-treated viscose fabrics to resist the microorganism's growth. On the contrary, a unique ability to stop growth of these microorganisms on the viscose fabric pretreated with 3-bromopropionic acid followed by treatment with nanometal oxides is also observed (El-Sayeed et al. 2015). It is also obvious that the ability of nanometal modified viscose fabrics to reduce the microbial growth is in the following order: zinc oxide > aluminium oxide > titanium (IV) oxide. The ability of nanometal oxide to decrease the number of viable microorganism colonies could be attributed to their environmental stress (reactive oxygen species level) on the bacterial cell wall, which may breakdown the

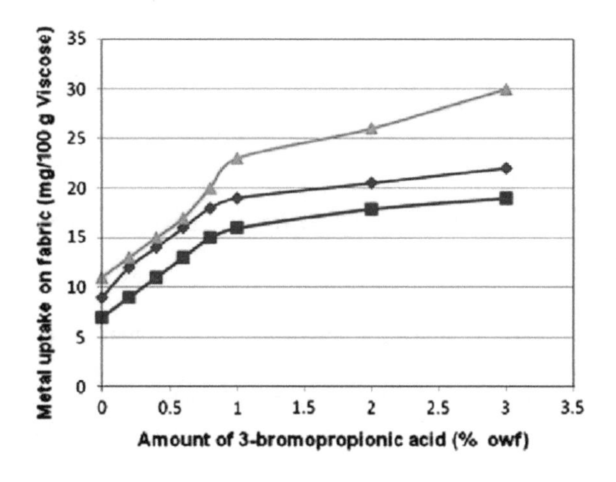

Fig. 30 Effect of concentration of 3-bromopropionic acid and nanometal oxides on antimicrobial activity of the viscose fabric

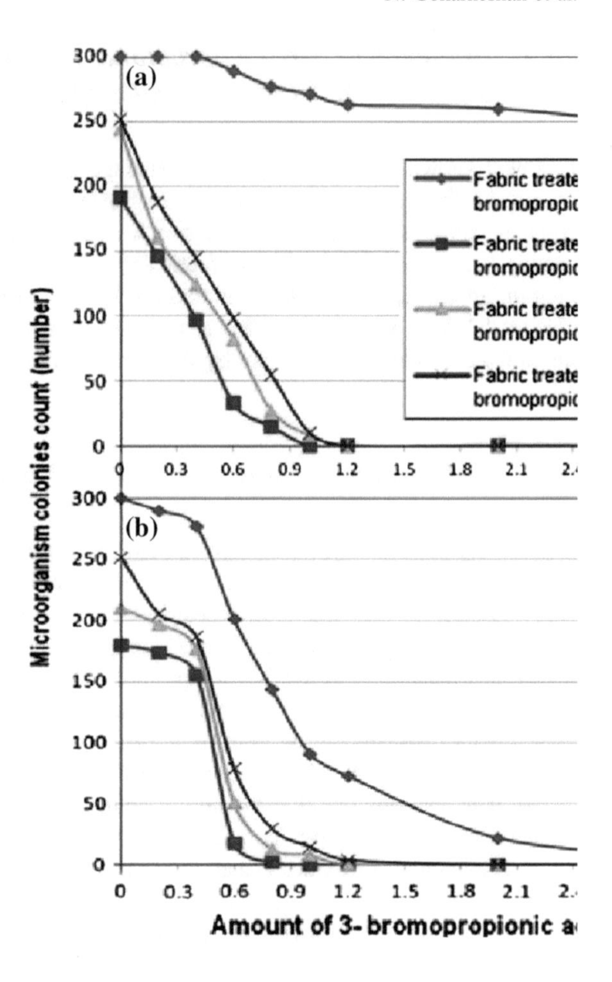

Fig. 31 Effect of 3-bromopropionic acid and nanometal oxides on viscose fabric (**a**) antibacterial activity of viscose fabric (**b**) and antifungal activity of viscose fabric (El-Sayeed et al. 2015)

cell wall and the outer membrane. It permits cell contents to leak out and nanometal oxide to enter, leading to cell irregularities and depressions (Okay 2010). It is also evident that the antifungal activity of 3-bromopropionic acid is superior compared to its antibacterial activity.

10.8 Durability of Antimicrobial Activity of Treated Viscose Fabric

In order to study the resistance to microbial growth, viscose fabrics have been treated with nanometal oxides and also viscose fabrics have been pretreated with 3-bromopropionic acid and/or nanometal oxides which have been exposed to 30 wash cycles. The viscose fabrics treated with 3-bromopropionic acid and after treated with nanometal oxides exhibit superior reduction in the microorganism's growth, in comparison with those fabrics treated only with nanometal oxides (El-Sayeed et al. 2015). This verdict could be attributed to the higher concentration of nanometal oxides on the fabrics treated with 3-bromopropionic acid. The carboxylic groups added to the fabrics are not only acting as attracting groups for the nanometal oxides, but also fixing them on the fabric, which could explain the ability of the viscose fabrics, treated with 3-bromopropionic acid and nanometal oxides, to maintain their high antimicrobial activity even after 30 wash cycles.

10.9 TEM Investigation

The morphological study of the untreated and treated viscose fabrics shows no clear difference in the morphological structure of the viscose fabrics after treatment with 3-bromopropionic acid and nanometal oxides (Fig. 32). It arises due to the ability of the nanoparticles to form a transparent layer on the fabric surface.

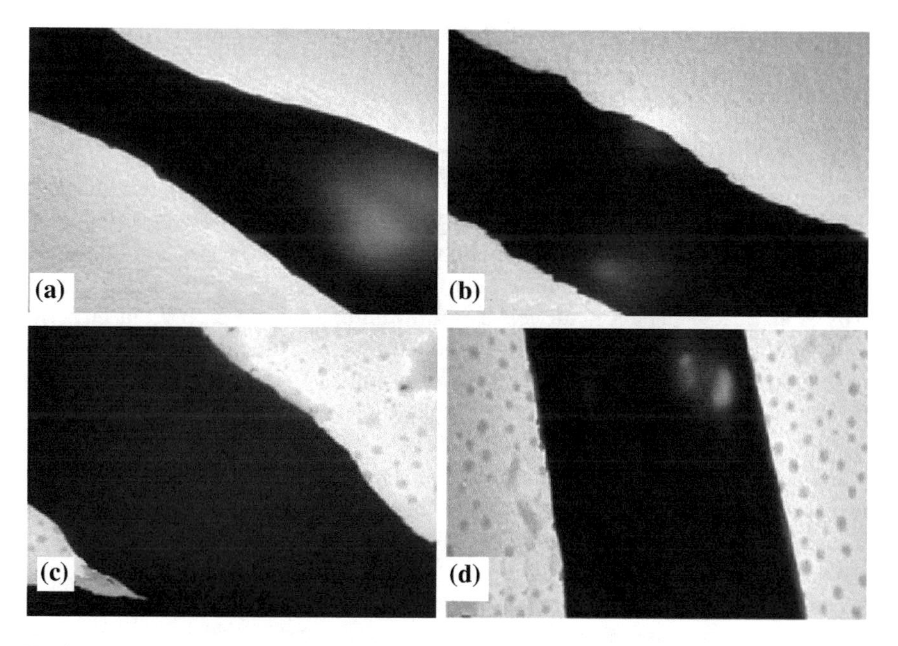

Fig. 32 TEM of untreated and pretreated viscose fabrics with 3-bromopropionic acid and nanometal oxides **a** untreated viscose, **b** pretreated viscose + Zn, **c** pretreated viscose + Al and **d** pretreated viscose + Ti]. (El-Sayeed et al. 2015)

11 Finishes with Stimuli-Responsive Nanohydrogels

11.1 Overview

Cotton fabric properties modified with nanoparticles of PNCS hydrogel reveal that they have acquired new smart responsiveness against pH and temperature. The thickness of modified fabrics does not exhibit significant difference with control samples, owing to small size of nanoparticles. There is no crease resistant effect on smart textiles, due to the BTCA, which links the hydrogel nanoparticle to the fabric. But, due to the involvement of the hydroxyl group in cross-linking process, the CRA of modified fabric has increased in comparison with control sample. Water vapour transition, air permeability, and wicking of smart modified cotton fabric are the main properties studied for determining fabric comfort (Bashari et al. 2015). Surface modification systems containing pH and temperature dual-responsive nanohydrogels due to the small size of particles show no negative effects on mentioned comfort parameters and the wicking of modified fabric is increased especially at the temperature above LCST (>32 °C). Ultimately, the findings confirm that the controlled contraction or expansion of the dual-responsive nanoparticles confers on the textile material smart properties, particularly with liquid management, and the functional material obtained reacts satisfactorily to the changes in ambient conditions as well as maintains the individual comfort. There are good prospects of using such smart cotton as an advanced material in sports, medical textiles, and other technical areas of applications.

11.2 Related Aspects

Hydrogels or also known as hydrophilic gels are cross-linked polymeric materials capable of absorption of lot of water without dissolving. The properties of softness, smartness, and the capacity to store water render them suitable for a number of uses (Roy et al. 2010). Certain hydrogels can show abrupt volume changes in response to external stimuli like temperature, pH, electric and magnetic fields, solvent quality, and light (Jocic et al. 2009). In recent years, more research developments in the functional finishing of textiles by stimuli-responsive polymeric systems are being made. Researchers have proven that they could apply submicron hydrogels on textiles in very thin layers. More surface-to-weight ratios of nanohydrogels than microhydrogels or bulk hydrogels make them more sensitive (Hu et al. 2012). Through this new finishing method, the new smart textile can be created, containing fibres that maintains conventional properties (flexibility, mechanical strength, and wearing comfort) but shows advanced functionalities and/or environmental responsiveness because of the surface modification of textile with a very thin layer of responsive hydrogel. The general concept of smart textile refers to textile structure that can sense and construe the stimuli in their environment and respond appropriately. However, they should not be confused with multifunctional or high-performance textiles that are non-active materials with special properties (Bashari et al. 2013). Stimuli-responsive hydrogels can be grafted onto the surfaces of cotton (CO), polypropylene (PP), and polyester (PET) fabrics by using different techniques (Bashari et al. 2013). The most often studied methods are preactivation of textile by chemical (cationization and anionization) or physical (air, N_2, Ar-plasma, or γ-irradiation) techniques or by using cross-linking agents (glutaraldehyde and BTCA) for applying hydrogel particle on non-activated textile substrate (Matsukuma et al. 2006). The focus has been on study of the conventional properties (particularly physiological comfort properties) of cotton fabric after applying the thin layer of pH-temperature dual-responsive PNIPAAm/chitosan nanohydrogels.

11.3 Technical Details

100 % desized cotton fabric has been used. The chemicals used include NIPAAm monomer (99 % pure stabilized), chitosan (medium molecular weight, viscosity 1 % solution in 1 % acetic acid, 200–800 cps), *N,N*-methylene bisacrylamide (MBA) and ammonium persulphate (APS), BTCA, sodium hypophosphite (SHP), non-ionic surfactant Adrasil HP (P-836), and *N,N,N′,N′*-tetra methyl ethylene diamine (TEMED). Methylene blue (MB) and other chemicals were analytically graded and used without further purification. The responsiveness of modified cotton fabric with PNCS nanogels against pH and temperature has been proved in earlier work. The water uptake (WU) and water retention capacity (WRC) of modified cotton fabric were investigated for evaluation of smart property of textile against two above-mentioned stimuli (Aguilar et al. 2007). The effect of surface modification on thickness of cotton fabric was assessed. The crease recovery angle of smart fabrics was measured. The yellowness index (YI) of modified fabric was measured. In order to study the stimuli-responsive finishing effect on physiological comfort parameters of cotton fabric, the water vapour transmission rate (WVT), air permeability, and vertical wicking of fabric were assessed. The water vapour transmission (WVT) of cotton fabrics was measured under two different conditions (25 and 40 °C and relative humidity 65 %). The lower critical solution temperature (LCST) of the temperature-responsive component of PNCS nanogels is found at around 32 °C, and hence two temperature conditions below and above the LCST have been chosen (Kittinaovarut 1998a). The rate of water vapour transmission has been expressed in g/m^2 for 24 h.

11.4 SEM Characterization

The incorporation of PNCS microgel onto the cotton surface has been confirmed by SEM (Fig. 33). The WU and WRC results confirm that due to the thermoresponsiveness of PNIPAAm and pH-responsiveness of chitosan, the smart fabric absorbs more water at temperature below LCST (∼32 °C) and with acidic pH (less than pK_b chitosan ∼6.5), while it absorbs less water when temperature is above LCST and pH is basic. Below LCST, the polymer chains, because of predomination of hydrogen bonding, are hydrophilic, whereas a phase separation occurs above the LCST due to predomination of hydrophobic interactions (Brojeswari et al. 2007). At acidic pH, amino groups of chitosan are protonated, and therefore PNIPAAm's negative charges (due to the polymerization procedure with APS, the PNIPAAm nanoparticles have negative charges) are neutralized by chitosan. Greater hydrophobicity than in their charged state results from the counterbalance of nanoparticles, where they are maintained hydrated and stable.

11.5 Influence of Nanohydrogel Finishing on Fabric Thickness

The influence of smart finishing on thickness of modified cotton fabrics has been evaluated. The samples 1–3 represent the cotton fabrics modified with 4, 6, and 8 (% owf) PNCS and proper amounts of the BTCA, SHP, and control sample in the cotton fabric without any hydrogel finishing on the surface (Bashari et al. 2015). The increase in thickness of the modified fabrics in comparison with control fabric indicates the presence of the smart nanogel system on fabric. For the thickness measured at 30 °C (below LCST of PNCS nanogel) and at around standard relative humidity of the environment, the effect of hydrogel presence on fabric thickness at

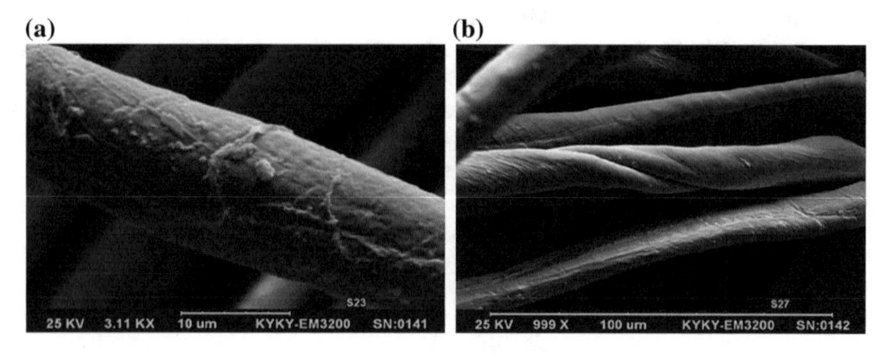

Fig. 33 SEM micrographs of **a** cotton fabric treated with PNCS nanoparticles, and **b** control fabric (Bashari et al. 2015)

temperature below LCST is more than that measured at the temperature above LSCT. But there is no increase beyond 5 % in the thickness of fabric samples.

11.6 Influence of Nanohydrogel Finishing on Fabric CRA

The CRA data of modified and reference fabrics in warp and weft directions have been obtained. The findings reveal that all modified samples have a little larger CRA than control sample. This is due to the involvement of the hydroxyl groups of the cellulose chains in cross-linking reaction with PNCS nanogel and BTCA and consequently the decrease in cellulose chain mobility to create wrinkles (Bashari et al. 2015). Since crease recovery angle of modified fabrics in comparison with control does not show significant changes, it cannot be said that the modified fabrics with nanogel have anticreasing property.

11.7 Influence of Nanohydrogel Finishing on Fabric Yellowness Index

The yellowness indices of the modified and control cotton fabrics were have been determined.

The obtained values reveal that yellowness of cotton fabrics has increased by almost 0.5. Even though visual inspection of the cotton samples does not show severe differences spectrophotometrically, it is proven otherwise. Excessive time and temperature of cotton fabric curing may scorch cellulosic fabrics and cause them to become yellow. A temperature of 160–180 °C is normally used as the curing temperature for formaldehyde-free durable press agents, such as BTCA, three-propane tricarboxylic acid (TCA), and citric acid (CA) (Bashari et al. 2015). The observed yellowing of the fabric finished with CA combinations is found more severe than that of fabric finished with BTCA combinations. The short duration of curing and the use of BTCA as a cross-linking agent result in modification of smart cotton fabrics with least yellowness index.

11.8 Influence of Nanohydrogel Finishing on Fabric WVT

The water vapour transmission rates (WVT) of modified and control cotton fabrics are depicted in Fig. 34. There are three modes of water vapour transmission (WVT) through the fabric (Kittinaovarut 1998b).

(i) Simple diffusion through distances between yarns in fabric structure.
(ii) Capillary transferring from the fibres.
(iii) Diffusion by each fibre.

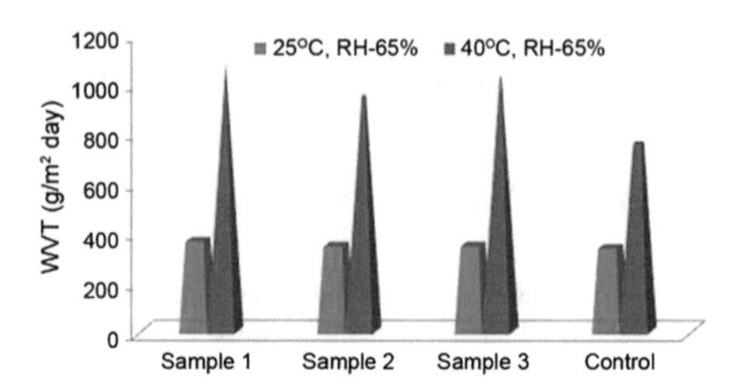

Fig. 34 WVT rates of modified and control cotton fabric [RH 65 %, 25 °C (<LCST) and 40 °C (>LCST)] (Bashari et al. 2015)

The transfer rates of water vapour at 40 °C are more than 25 °C, as revealed by the WVT results of modified and control cotton fabrics.

At temperatures below the LCST of PNCS nanogels, the hydrogel nanoparticles are in swollen state and at temperatures above the LCST, the nanoparticles are shrunk, and this leads to increase and decrease in moisture content of the nanoparticles on the surface of the fabric respectively. At 40 °C, the shrinking of nanoparticles allows the water vapour to pass through the fabric easier than under the condition when nanogels are in swelling state. There is no significant difference found between the WVT of modified and control samples at 25 °C (Bashari et al. 2015). Although the swelling of nanogels at this temperature causes the transmission of less water vapour through the spaces between the fibres (Mechanism 1), with increasing the moisture content of particles, the probability of water vapour transmission through the surface diffusion and capillary methods (Mechanisms 2 and 3) increases. The two effects compensate each other and hence the water vapour transmission rate of the modified textile fabrics at 25 °C is retained near the WVT of control fabric.

11.9 Influence of Nanohydrogel Finishing on Air Permeability of Fabric

The air permeability results of modified and control cotton fabrics is depicted in Fig. 35. When compared with the control sample, the air permeability of modified fabrics at laboratory conditions (30 °C, RH 60 %) has slightly decreased. It may be related to the inherent swelling property of hydrogel nanoparticle under this condition (Bashari et al. 2015). On the other hand, as the ambient temperature is below the LCST of PNCS nanoparticles, hydrogel nanoparticles are in swelling form and therefore some parts of spaces between fibres are blocking and the passage of air

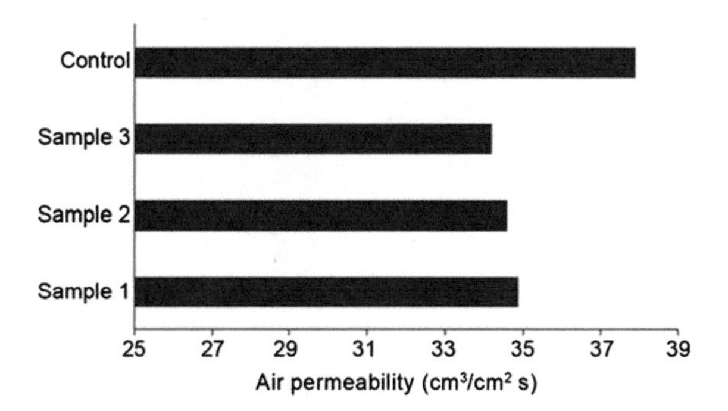

Fig. 35 Air permeability of modified and control cotton fabric [RH 65 %, 30 °C] (Bashari et al. 2015)

through the fabric is reduced. But since the hydrogel is in nanoparticle form and amount of this surface modifying system on fabric is not more than 8 % (owf), reduction in amount of air permeability through the modified fabrics in comparison with control fabric is not so marked.

11.10 Influence of Nanohydrogel Finishing on Fabric Vertical Wicking

Figure 36 depicts the vertical wicking behaviour of modified and control cotton fabrics. The following factors influence capillary and the liquid rises in the length of capillary tubes

(a) Contact angle (θ)
(b) Surface tension (ρ)
(c) Diameter of the capillary tube (R).

Hydrophilic textiles more readily absorb water than hydrophobic textiles. More surface area is wetted with the increase in contact angle when the hydrophilicity of fabric is increased. Hence, the wetting time in modified cotton fabric with hydrogel

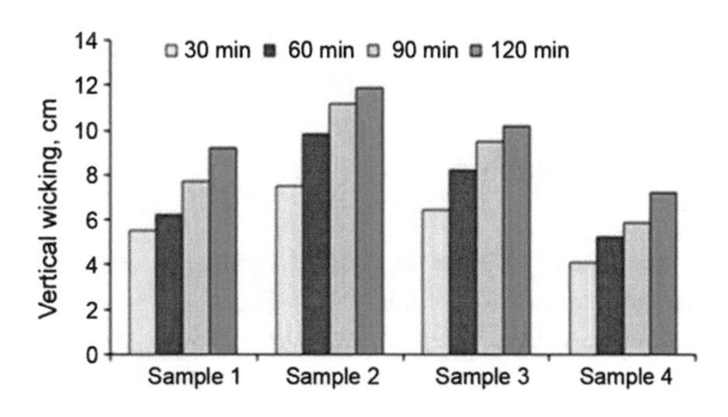

Fig. 36 Vertical wicking of modified and control cotton fabric [RH 65 %, 30 °C] (Bashari et al. 2015)

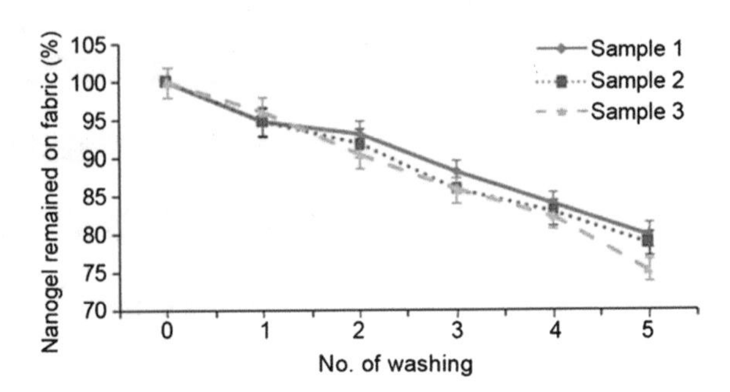

Fig. 37 Wash fastness of modified cotton fabric with smart nanogels after 5 times washing (Bashari et al. 2015)

systems are less than in control fabric due to the hydrophilic nature of hydrogel on the fabric surface. In terms of surface tension's role on fabric wicking, considering that potassium permanganate solution was used in all tests, the surface tension is found similar for both modified and control fabrics. The diameter of capillary tubes is another factor that affects the vertical wicking of fabric. The swelling of the hydrogel system present on the fabric surface can lead to changes in the diameter of the capillary-like tubes (open pores between the fibres) in the fabric (Bashari et al. 2015). The fluid speed increases with decreasing the diameter of the capillary tube. As the fabric with a plain weave is used in this study, the structure of fabric is relatively open; closing of pores between the fibres can lead to a significant reduction in the wicking of the fabric. The wicking of modified smart fabric is increased as shown in Fig. 36. Thus, due to nanogel swelling, the increase in hydrophilicity of fabric surface and decrease in diameter of open pores of fabric structure result in increase in wicking parameter.

11.11 Assessment of Washing Fastness of Modified Fabrics

The three modified fabrics have been washed 5 times so as to check the amount of the nanogel finishing system that remains on fabric, as per stipulated conditions (Fig. 37). After five washing process, about 75–80 % of the initial hydrogel system has been remained on the fabric surface. This represents the acceptable washing fastness of smart cotton fabric.

12 Conclusion

It has been possible to modify cotton fabrics into garments for special needs through application of nanosilver finishes. Various properties were affected such as air permeability, water vapour permeability, wrinkle recovery angle, and bending rigidity. Increasing the concentration of the nanosolution beyond 500 ppm marked a decline in most of the chemical and physical properties. Finishing has been done with herbal extracts on denim fabric. The durability of the finishing has been enhanced through microencapsulation and nano encapsulation techniques. Antimicrobial effect is present even for the washed fabrics against the standard strains. Application of nanochitosan on cotton fabrics improves appearance and handle, fibre strength, absorbency, washing fastness and wrinkle recovery. Nanochitosan together with nanosilver treatment shows enhanced antibacterial activity. Disperse dye reduced to nanosize of 23 nm has been exposed to ultrasound, used in printing paste, and applied to polyester fabric. The prints show very good fastness properties and no differences can be observed from prints of untreated dye with ultrasound. Studies on nanopolysiloxane finishing on jute-blended fabrics shows improvement in handle, flexural rigidity, and crease recovery. Combination of nano + micropolysiloxane finishing has potential to improve the handle properties of jute-blended fabric. Studies on viscose fabrics treated with 3-bromopropionic acid and nanometal oxides are found able to maintain high antimicrobial activity even after 30 wash cycles. Zinc oxide gives highest antimicrobial activity followed by aluminium oxide and titanium dioxide. Cotton fabrics treated with nanoparticles of PCNS hydrogel show that these fabrics have acquired new smart responsiveness against pH and temperature. The use of natural materials in nanotextile finishing promotes ecofriendliness in processing, which is an important aspect from sustainability point of view. Also marked improvement in the properties and wider areas of applications of the new types of nanotextile finishes render these approaches more viable from the commercial point of view. Thus they hold the key in improvement of sustainability in the nanotextile finishing processes.

References

Achwal, W. B. (2000). UV protection by textiles. *Colourage,* (4), 50.

Achwal, W. B. (2003). Chitosan and its derivatives for textile finishing. *Colourage, 50*(8), 51–76.

Aguilar, M. R., Elvira, C., Gallardo, A., Vazquez, B., & Roman, J. S. (2007). Smart polymers and their applications as biomaterials, III biomaterials. *Topics in Tissue Engineering,* 1.

Ahmed, W. Y. W., & Loman, M. (1996). *Journal of the Society of Dyers and Colourists, 112*, 245.

American Association of Textile Chemists and Colorists. (2003). AATCC Test Method 66-2003. New York, USA: American Association of Textile Chemists and Colorists.

Ammayappan, L., & Moses, J. J. (2010). Functional finishing of jute. *Current Chemical Research, 1*(1), 19.

Ammayappan, L., Moses, J., Asok, S., Raja, A. S. M., & Jimmy, K. C. L. (2011). Performance properties of multi-functional finishes on the enzyme-pretreated wool/cotton blend fabrics. *Textile Color Finishing, 23*(1), 1.

Ammayappan, L., Nayak, L. K., & Ray, D. P. (2013). Value addition of jute textiles: present status and future perspectives. In K. K. Satapathy & P. K. Ganguly (Eds.), *Diversification of the jute and allied fiber: Some recent developments* (pp. 203–224). Kolkata: NIRJAFT.

Anna, N., Jonas, E., Bengt, H., Pernilla, W., & IFP Research AB. (2007). *The Nordic Textile Journal*. ISSN 1404-2487, 90–99.

Ansari, N., & Maleki, V. (2007). *Principles and theories of physical testing of fibers, yarns and fabrics* (pp. 159–218). Iran: Jihad Amirkabir University Publication Center.

Bagherzadeh, R., Montazer, M., Latifi, M., Sheikhzadeh, M., & Sattari, M. (2007). *Fibers and Polymers, 8*, 386.

Bajaj, P. (2002). Finishing of textile materials. *Journal of Applied Polymer Science, 83*(202), 631.

Barari, M., Majidi, R. F., & Madani, M. (2009). Preparation of nanocapsules via emulsifier-free miniemulsion polymerization. *Nanoscience and Nanotechnology, 9*, 4348.

Bashari, A., Hemmati Nejad, N., & Pourjavadi, A. (2015). Effect of stimuli-responsive nano hydrogel finishing on cotton fabric properties. *Indian Journal of Fibre & Textile Research, 40*, 431.

Bashari, A., Hemmati Nejad, N., & Pourjavadi, A. (2013a). Applications of stimuli responsive hydrogels: a textile engineering approach. *Journal of the Textile Institute, 104*(11), 1145.

Bashari, A., Hemmati Nejad, N., & Pourjavadi, A. (2013b). Surface modification of cotton fabric with dual-responsive PNIPAAm/chitosan nano hydrogel. *Polymers for Advanced Technologies, 24*(9), 797.

Bendak, A., Raslan, W., & Salama, M. (2008). Treatment of wool with metal salts and their effects on its properties. *Journal of Natural Fibers, 5*(3), 251.

Benita, S. (ed.). (1996). Microencapsulation methods and industrial application. *Drugs and the Pharmaceutical Sciences,* (2nd ed., p. 158).

Bhoomika, G. R., Ramesh, G. K., & Anita, M. A. (2007). Phyto-pharmacology of Achyranthes asera: a review. *Pharmacognosy Review, 1*, 143–149.

Boonyo, W., Junginger, H. E., Waranuch, N., Polnok, A., & Pitaksuteepong, T. (2008). *Journal of Metallurgy, Materials and Minerals, 18*(2), 59.

British Standards. (2012). BS 3356:1990. London: British Standards Institution.

Broadbent, A. D. (2001). *Basic principles of textile coloration* (p. 322). England: Society of Dyers & Colorists, Thanet Press Ltd.

Brojeswari, D., Apurba, D., Kothari, V. K., Fangueiro, R., & de Araújo, M. (2007). Moisture transmission through textiles, Part I: Processes in moisture transmission and the factors at play. *AUTEX Research Journal, 7*(3), 194.

Campus, F., Bonhote, P., Grätzel, M., Heinen, S., & Walder, L. (1999). *Solar Energy Materials and Solar Cells, 56*, 281.

Chattopadhyay, D. P., & Inamdar, M. S. (2009). Studies on the properties of chitosan treated cotton fabric. *Asian Dyer, 6*(5), 47–53.

Chattopadhyay, D. P., & Inamdar, M. S. (2010). Aqueous behaviour of chitosan. *International Journal of Polymer Science, 2010*, 1.

Chattopadhyay, D., & Inamdar, M. S. (2013). Improvement in properties of cotton fabric through synthesized nano-chitosan application. *Indian Journal of Fibres and Textile Research, 38*, 14.

Chattopadhyay, D. P., & Patel, B. H. (2009). Improvement in physical and dyeing properties of natural fibres through pretreatment with silver nanoparticles. *Indian Journal of Fibre & Textile Research, 34*, 368.

Daoud, W. A., & Xin, J. H. (2004). Low temperature sol-gel processed photocatalytic titania coating. *Journal of Sol-Gel Science and Technology, 29*, 25.

Daoud, W. A., Xin, J. H., & Zhang, Y.-H. (2005). Surface functionalization of cellulose fibers with titanium dioxide nanoparticles and their combined bactericidal activities. *Surface Science, 599* (1), 69.

Dastjerdi, R., Mojtahedi, M., Shoshtari, A., & Khosroshahi, A. (2010). Investigating the production and properties of Ag/TiO2/PP antibacterial nanocomposite filament yarns. *Journal of the Textile Institute, 101*(3), 204.

Dastjerdi, R., Montazer, M., & Shahsavan, S. (2009). A new method to stabilize nanoparticles on textile surfaces. *Colloids and Surfaces A: Physicochemical and Engineering Aspects, 345*(3), 202.

Debnath, S., & Sengupta, S. (2009). Effect of linear density, twist and blend proportion on some physical properties of jute and hollow polyester blended yarn. *Indian Journal of Fibre & Textile Research, 34*(1), 11.

Debnath, S., Sengupta, S., & Singh, U. S. (2007a). Properties of jute and hollow-polyester blended bulked yarn. *Journal of Institution of Engineers (India) Textile Engineering Divison, 87*(1), 11.

Debnath, S., Sengupta, S., & Singh, U. S. (2007b). Comparative study on the physical properties of jute, jute-viscose and jute-polyester (hollow) blended yarns. *Journal of Institution of Engineers (India) Textile Engineering Divison, 88*(8), 5.

Debnath, S., Sengupta, S., & Singh, P. (2011). *Annual report 2010-11* (p. 17). Kolkata: NIRJAFT.

Dhananjeyan, M., Mielczarski, E., Thampi, K., Buffat, P., Bensimon, M., Kulik, A., et al. (2001). *Journal of Physical Chemistry B, 105*, 12046.

Du, W.-D., Niu, S.-S., Xu, Y.-L., Xu, Z.-R., & Fan, C.-L. (2009). Antibacterial activity of chitosan tripolyphosphate nanoparticles loaded with various metal ions. *Carbohydrate Polymer,* 385.

Ebrahim, F. F. S., & Mansour, O. S. M. (2013). Using Nanomaterials treatments to improve the performance characteristics of garment groups with special needs. *Journal of American Science, 9*(11), 126–131.

El-Molla, M. M., El-Khatib, E. M., El-Gammal, M. S., & Abdel-Fattah, S. H. (2011). Development of ecofriendly binders for pigment printing of all types of textile fabrics. *Indian Journal of Fibre & Textile Research, 36*(3), 266.

El-Sayed, A. A., Dorgham, S. M., & Kantouch, A. (2012). Application of reactive salicylanilide to viscose fabrics as antibacterial and antifungus finishing. *International Journal of Biological Macromolecules, 50*(1), 273.

El-Sayed, A. A., El Gabry, L., & Allam, O. (2010). Application of prepared waterborne polyurethane extended with chitosan to impart antibacterial properties to acrylic fabrics. *Journal of Materials Science: Materials in Medicine, 21*(2), 507.

El-Sayeed, A. A., Salama, M., Sohad, M. D., & Kantouch, A. (2015). Modification of viscose fabrics to impart permanent antimicrobial activity. *Indian Journal of Fibres and Textile Research, 40*, 25.

El-Tahlawy, K. F. (1999). Utilization of citric acid-chitosan-sodium hypophosphite system for effecting concurrent dyeing and finishing. *Colourage, 46*, 21.

Enescu, D. (2008). Use of chitosan in surface modification of textile materials. *Roumanian Biotechnological Letters, 13*(6), 4037.

Eom, S. I. (2001). Using chitosan as an antistatic finish for polyester fabric. *AATCC Review, 1*(3), 57.

Fu, G., Vary, P. S., & Lin, C.-T. (2005). Anatase TiO2 Nanocomposites for Antimicrobial Coatings. *Journal of Physical Chemistry B, 109*(18), 8889.

Fung, W., & Hardcastle, M. (2001). *Textiles in automation engineering* (p. 120). England: Woodhead Publications.

Gao, Y., & Cranston, R. (2008). Recent advances in antimicrobial treatments of textiles. *Textile Research Journal, 78*(1), 60.

Gaurav, K. (2005). Antimicrobial for textiles. *Colourage, 52*(9), 94.

Giri Dev, V. R., Neelkandan, R., Sudha, N., Shamugasundaram, O. L., & Nadaraj, R. N. (2005). A material with wider applications. *Textile Magazine,* 83.

Gokarneshan, N, Gopalakrishnan, P. P., & Jeyanthi, B. (2012). Influence of various nano finishes on antibacterial properties of fabrics. *ISRN Nanomaterials*, 1–8.

Gokarneshan, N., Gopalakrishnan, P. P., & Jeyanthi, B. (2013). *Nano finishing of textiles*. New Delhi: Abhishek Publications.

Gorensek, M., & Recel, P. (2007). Nanosilver functionalized cotton fabric. *Textile Research Journal, 77*(3), 138.

Harish Prashant, K. V., & Tharanathan, R. N. (2007). Chitin/chitosan: modifications and their unlimited application potential—an overview. *Trends in Food Science and Technology, 18*(3), 117.

Harrocks, A. R., & Anand, A. (2000). *Handbook of technical textiles* (pp. 18, 192). England: Woodhead Publications.

Hasebe, Y. (2001). *AATCC Review, 1*(11), 23.

Hasebe, Y., Kuwahara, K., & Tokunaga, S. (2001). *AATCC Review—American Association of Textile Chemists and Colorists, 1*, 23.

Hatiboglu, B. (2006). *Mechanical properties of individual polymeric micro and nanofibers using atomic force microscopy (AFM)*. Ph.D. Dissertation, North Carolina. Available from www.lib.ncsu.edu/theses/available/etd-07062006-135651/unrestricted/etd.pdf
Hirano, S. (2003). *Ullmann's encyclopedia of industrial chemistry* (Vol. 7, p. 679). Weinheim, Germany: Wiley-VCH.
Holme, I. (2002). Microencapsulation of herbal extracts for microbial resistance in cotton fabrics. *Textile Magazine, 4*, 13.
Hu, J., Meng, H., Li, G., & Ibekwe, S. I. (2012). *Smart Materials and Structures, 21*(5), 23.
Huang, K.-S., Sheu, Y.-R., & Chao, I.-C. (2009). Preparation and properties of nanochitosan. *Polymer-Plastics Technology and Engineering, 48*(12), 1239.
Huang, W., Xing, Y., Yu, Y., Shang, S., & Dai, J. (2011). Enhanced washing durability of hydrophobic coating on cellulose fabric using polycarboxylic acids. *Applied Surface Science, 257*, 4443.
Huanga, Z. M., Zhangb, Y. Z., Kotakic, M., & Ramakrishnab, S. (2003). A review on polymer nanofibers by electrospinning and their applications in nanocomposites. *Composites Science and Technology, 63*(1), 2223–2253.
Inamdar, M. S., & Chattopadhyay, D. P. (2006). Chitosan and its versatile applications in textile processing. *Man Made Textiles in India, 49*(6), 212.
Jocic, D., Tourrette, A., Glampedaki, P., & Warmoeskerken, M. M. C. G. (2009). Application of temperature and pH responsive micro hydrogels for functional finishing of cotton fabric. *Materials Technology: Advanced Performance Materials, 24*, 14–23.
Kantouch, A., & El-Sayed, A. A. (2008). Polyvinyl pyridine metal complex as permanent antimicrobial finishing for viscose fabric. *International Journal of Biological Macromolecules, 43*, 451.
Kantouch, A., El-Sayed, A. A., Salama, M., El-Kheir, A. A., & Mowafi, S. (2013). Salicylic acid and some of its derivatives as antibacterial agents for viscose fabric. *International Journal of Biological Macromolecules, 62*, 603.
Kean, T., Roth, S., & Thanou, M. (2005). Trimethylated chitosans as non-viral gene delivery vectors: cytotoxicity and transfection efficiency. *Journal of Controlled Release, 103*(3), 643.
Ki, H. Y., Kim, J. H., Kwon, S. C., & Jeong, S. H. (2007). A study on multifunctional wool textiles treated with nano-sized silver. *Journal of Materials Science, 42*, 8020.
Kittinaovarut, S. (1998a). Polymerization-crosslinking fabric finishing, with pad-dry-cure, using nonformaldehyde BTCA/IA/AA combinations to impart durable press properties in cotton fabric. In *Near environments*. Radford: Radford University.
Kittinaovarut, S. (1998b). Cure, using nonformaldehyde BTCA/IA/AA combinations to impart durable press properties. In *Cotton fabric*. Radford: Radford University.
Knittel, D., & Schollmeyer, E. (2002). Permanent finishing of cotton with ionic carbohydrates and analysis of thin layers obtained. *Melliand Textilber, 83*, 15.
Kundu, B. C. (1956). Jute—world's foremost bast fibre. Botany, agronomy, diseases and pests. *Economic Botany, 10*, 103.
Lachapelle, J. M., & Maibach, H. I. (2009). *Patch testing and prick testing: A practical guide official publication of the ICDRG*. Berlin, Germany: Springer, 60.
Ladhari, N., Baouab, M., Ben Dekhil, A., Bakhrouf, A., & Niquette, P. (2007). Antibacterial activity of quaternary ammonium salt grafted cotton. *Journal of the Textile Institute, 98*, 209.
Lakshmanan, A., Debnath, S., & Sengupta, S. (2014). Effect of nano-polysiloxane based finishing on handle properties of jute blended fabric, *Indian Journal of Fibres and Textile Research, 39*, 425.
Lee, K. W., Chung, Y. S., & Kim, J. P. (2001). Effect of ultrasonic on disperse dye particle size. *Textile Research Journal, 71*(5), 395.
Lee, H. J., & Jeong, S. H. (2005). Bacteriostasis and skin innoxiousness of nanosize silver colloids on textile fabrics. *Textile Research Journal, 75*(7), 551.
Lee, K. W., & Kim, J. P. (2001). Effect of ultrasonic on disperse dye particle size. *Textile Research Journal, 71*(5), 395.

Leslie, W. C. M. (1994). *Textile printing* (2nd ed., pp. 174, 175). London, UK: Society of Dyers & Colorist.

Li, Q., Chen, S. L., & Jiang, W. C. (2007). Durability of nano ZnO antibacterial cotton fabric to sweat. *Journal of Applied Polymer Science, 103*(1), 412.

Liu, J.-K., Yang, X.-H., & Tian, X.-G. (2008). Preparation of silver/hydroxyapatite nanocomposite spheres. *Powder Technology, 184*(1), 21.

Lopez-Leon, T., Carvalho, E. L. S., Seijo, B., Ortega-Vinuesa, J. L., & Bastos-Gonzalez, D. (2005). Physicochemical characterization of chitosan nanoparticles: electrokinetic and stability behavior. *Journal of Colloid and Interface Science, 283*(2), 344.

Loretz, B., & Bernkop, S. (2006). In vitro evaluation of chitosan-EDTA conjugate polyplexes as a nanoparticulate gene delivery system. *AAPS Journal, 8*(4), 756–764. (Art. no. 85).

Matsukuma, D., Yamamoto, K., & Aoyagi, T. (2006). Stimuli-responsive properties of N-isopropylacrylamide-based ultrathin hydrogel films prepared by photo-cross-linking. *Langmuir, 22*(13), 5911.

Mekewi, M., El-Sayed, A. A., Amin, M., & Said, H. I. (2012). Imparting permanent antimicrobial onto viscose and acrylic fabrics. *International Journal of Biological Macromolecules, 50*, 1055.

Montazer, M., & Afjeh, M. G. (2007). Simultaneous X-linking and antimicrobial finishing of cotton fabric. *Journal of Applied Polymer Science, 103*(1), 178–185.

Muzzarelli, R. A. A. (1996). Chitin chemistry. In J. C. Salamone (Ed.), *The polymeric materials encyclopedia* (pp. 312–314). Boca Raton, FL, USA. CRC Press Inc.

Nahed, S. E. A., & El-Shishtawy, R. M. (2010). *Journal of Material Science*, *45*, 1143.

Nakashima, T., Sakagami, Y., Ito, H., & Matsuo, M. (2001). *Textile Research Journal, 71*, 688.

Natarajan, V. (2002). Azadirachta indica in the treatment of dermatophytosis. *Journal of Ecobiology, 14*(3), 201.

No, H. K., & Meyers, S. P. (1995). Preparation and characterization of chitin and chitosan: a review. *Journal of Aquatic Food Product Technology, 4*(2), 27.

Okay, O. (2010). General properties of hydrogels. In G. Gerlach & K.-F. Arndt (Eds.), *Hydrogel sensors and actuators* (pp. 1–14). Berlin: Springer.

Oktem, T. (2003). Surface treatment of cotton fabrics with chitosan. *Coloration Technology, 119* (4), 241.

Osman, H., & Khairy, M. (2013). Optimization of polyester printing with disperse dye nanoparticles. *Indian Journal of Fibres and Textile Research, 38*, 202.

Parikh, D., Fink, T., Rajasekharan, K., Sachinvala, N., Sawhney, A., Calamari, T. (2005). Comparative study of synergistic effects of antibiotics with triangular shaped silver nanoparticles, synthesized using UV-light irradiation, on Staphylococcus aureus and Pseudomonas aeruginosa. *Textile Research Journal, 75*, 134.

Parthasarathi, V. (2008). *Nano technology adds value to textile finishing*. Available from www.indiantextilejournal.com/articles, February 20, 2008.

Patel, J. K., & Jivani, N. P. (2009). Chitosan based nanoparticles in drug delivery. *International Journal of Pharmaceutical Sciences Nanotechnology, 2*(2), 517.

Perelshtein, I., Applerot, G., Perkas, N., Guibert, G., Mikhailov, S., & Gedanken, A. (2008). Sonochemical coating of silver nanoparticles on textile fabrics (nylon, polyester and cotton) and their antibacterial activity. *Nanotechnology, 19*, 245705.

Pujari, M. M., Kulkarni, M. S., & Kadole, P. V. (2010). Bombay to Goa Journey of the denim. *Textile Review, 5*(11), 7–9.

Ren, X., Kou, L., Kocer, H. B., Zhu, C., Worley, S., Broughton, R., et al. (2008). *Colloids and Surfaces A: Physicochemical and Engineering Aspects, 317*, 711.

Rowell, R. M., Stout, H. P. (1998). Jute and kenaf. In M. Lewin & E. M. Pearce (Eds.), *Handbook of fiber chemistry* (pp. 465–504). New York: Marcel Dekker.

Roy, D., Cambre, J. N., & Sumerlin, B. S. (2010). Future perspectives and recent advances in stimusli-responsive materials. *Progress in Polymer Science, 35*(1–2), 278.

Salama, M., Bendak, A., & Moller, M. (2011). *Industria Textila, 62*, 320.

Saligram, A. N., Shukla, S. R., & Mathur, M. (1993). Physico chemical aspects of textile coloration. *Journal of Society of Dyers and Colorists, 109*, 263.

Sanjay Vishwakarma, *Text Rev*, November (2010) 13.

Sathianarayanan, M. P., Bhat, M. V., Kokale, S. S., & Walunj, V. E. (2011). *Indian Journal of Fibre & Textile Research, 36*, 234.

Saville, B. P. (1999). *The physical testing of textiles*. London: Woodhead Publication Limited.

Schindler, W. D., & Hauser, P. J. (2004). *Chemical finishing of textiles* (p. 51). Cambridge, England: Woodhead Publishing Limited.

Shilpa, U. N. (2004). *Journal of Text Association, 65*, 219.

Shin, Y., Yoo, D. I., & Min, K. (1999). Durable antibacterial finish on cotton fabric by using chitosan-based polymeric core-shell particles. *Journal of Applied Polymer Science, 74*, 2911.

Srikanth, S. (2010). *Apparel Views*, 60.

Steele, R. (1962). Some fabric properties and their relation to crease proofing effects. *Journal of the Textile Institute, 53*(1), 7.

Subhash, A., & Sarkar Ajoy, K. (2010). *Colourage, 57*, 57.

Sumithra, M., & Vasugi Raaja, N. (2012). Micro-encapsulation and nano-encapsulation of denim Fabrics with herbal extracts. *Indian Journal of Fibres and Textile Research, 37*, 321.

Tan, E. P. S., & Lim, C. T. (2006). Mechanical characterization of nanofibers—A review. *Composites Science and Technology, 66*(1), 1102–1111.

Tatiana, L., Jose, N., & Fernando, O. (2008). *Proceedings of 4th International Textile, Clothing & Design Conference—Magic World of Textiles, Dubrovnik, Croatia*, December 2008.

Thies, C. (2005). *Microencapsulation* (4th ed., pp. 628–651). New York: Wiley.

Thilagavathi, G., Bala, S. K., & Kannaian, T. (2007). Microencapsulation of herbal extracts for microbial resistance in healthcare textiles. *Indian Journal of Fibre & Textile Research, 32*, 351.

Tiwari, S. K., & Gharia, M. M. (2003). Characterization of chitosan pastes and their application in textile printing. *AATCC Review, 3*(4), 17.

Tong, Y., Tian, M., Xu, R., Hu, W., Yu, L., & Zhang, L. (2003). *Fuhe Cailiao Xuebao (Acta Materiae Compositae Sinica (China), 20*, 88.

Trapani, A., Sitterberg, J., Bakowsky, U., & Kissel, T. (2009). The potential of glycol chitosan nanoparticles as carrier for low water soluble drugs. *International Journal of Pharmaceutics, 375*, 97.

Wong, Y. W. H., Yuen, C. W. M., Leung, M. Y. S., Ku, S. K. A., & Lam, H. L. I. (2006a). *AUTEX Research Journal, 6*(1), 1–8.

Wong, Y., Yuen, C., Leung, M., Ku, S., & Lam, H. (2006b). Selected applications of nano technology in textiles. *AUTEX Research Journal, 6*, 1.

Yang, J. M., Lin, H. T., Wu, T. H., & Chen, C. C. (2003). Wettability and antibacterial assessment of chitosan containing radiation-induced graft nonwoven fabric of polypropylene-g-acrylic acid. *Journal of Applied Polymer Science, 90*, 1331.

Yeo, S. Y., Lee, H. J., & Jeong, S. H. (2003). Antibacterial effect of nanosized silver colloidal solution on textile fabrics. *Journal Materials Science, 38*(10), 2199–2204.

Zhang, Z., Chen, L., Ji, J., Huang, V., & Chen, D. (2003). Antibacterial properties of cotton fabrics treated with chitosan. *Textile Research Journal, 73*(12), 1103.

Zhang, H., Wu, S., Tao, Y., Zang, L., & Su, Z. (2010). Preparation and in vitro characterization of chitosan nanoparticles. *Journal of Nanomaterials, 151*(2), 458–465.

Nanochemicals and Effluent Treatment in Textile Industries

P. Senthil Kumar, Abishek S. Narayan and Abhishek Dutta

Abstract The textile industry deals with the design and production of various fabrics with a web of processes intertwined to produce the final product. Among these, the dyeing and finishing processes in particular use large quantities of water and consequently lead to production of large volumes of wastewater. Colour, dissolved solids, toxic heavy metals, residual chlorine and other non-degradable organic materials are the pollutants of major concern present in effluent from textile industries. Advancements in nanotechnology have enabled us to explore the applications of nanochemicals for effluent treatment in textile industries. Nanochemicals have the desired properties required for pollutant and pathogen removal from wastewater by methods such as chemical oxidation, disinfection and photocatalysis. This chapter discusses the various pollutants present in wastewater effluent from textile industries and their sources, the present effluent standards and the application of nanochemicals for wastewater treatment.

Keywords Textile industry · Wastewater treatment technologies · Novel textile effluent treatment · Organic and inorganic impurity removal · Nanochemicals for wastewater treatment

1 Introduction

Textile industry is concerned with the design and production of textile fibres such as yarn, cotton, wool and silk. This industry plays a crucial part in providing the society with basic needs. It is also important in the economic perspective, providing employment and high industrial output. Especially developing countries have high economic contribution from textile industries.

P. Senthil Kumar (✉) · A.S. Narayan · A. Dutta
Department of Chemical Engineering, SSN College of Engineering,
Chennai 603110, India
e-mail: senthilchem8582@gmail.com

S.S. Muthu (ed.), *Textiles and Clothing Sustainability*,
Textile Science and Clothing Technology,
DOI 10.1007/978-981-10-2188-6_2

The textile industry spans a spectrum of processes. It can range from the small-scale cottage industries and spinning looms with low economic output, usually established in rural areas which are of domestic value, to the large-scale industries which involve dyeing and designing. The present-day industries use integrated mills which are used to convert different raw materials such as cotton and silk to different products using a unified facility. They also use modern machines with advanced technology such as carding machines for producing spun yarn and draw frames for combining two fibres.

In India, the textile industry has been prevalent for several centuries and has also been among the important contributors to the economic development of the country. This industry alone accounted for earnings through export worth US$41.4 billion in the fiscal year 2014–2015, an annual growth of 5.4 %, as released by the Cotton Textiles Export Promotion Council.

The textiles industry in India is tremendously diverse, with hand-spun and hand-woven textiles sectors in one end of the spectrum, and the high capital sophisticated mills sector in the other end as shown in Table 1. The biggest

Table 1 Types of textile industries in India

Type of textile industry	Units	2007–2008	2008–2009	2009–2010 (prov.)	2010–2011 (prov.)
Cotton/man-made fibre textile					
Mills	No.	1773	1830	1853	1940
Spinning mills (non-SSI)	No.	1597	1653	1673	1757
Composite mills (non-SSI)	No.	176	177	180	183
Spinning mills (SSI)	No.	1219	1247	1260	1333
Exclusive weaving mills (non-SSI)	No.	179	184	183	174
Powerloom units	Lakh No.	4.69	4.94	5.05	5.18
Capacity installed					
Spindles (SSI + non-SSI)	Million No.	39.07	41.34	42.04	47.58
Rotors (SSI + non-SSI)	Lakh No.	6.21	6.60	6.75	7.49
Looms (organized sector)	Lakh No.	0.71	0.71	0.71	0.66
Powerloom	Lakh No.	21.06	22.05	22.46	22.91
Handloom	Lakh No.	38.91	38.91	23.77#	23.77#
Man-made fibres	Million kilogram	1659	1763	1763	1765
Man-made filaments	Million kilogram	2101	2143	2188	2196
Worsted spindles (woollen)	Thousand No.	604	604	604	604
Non-worsted spindles (woollen)	Thousand No.	437	437	437	437

(continued)

Table 1 (continued)

Type of textile industry	Units	2007–2008	2008–2009	2009–2010 (prov.)	2010–2011 (prov.)
Production of fibres					
Raw cotton[a]	Lakh bales	307	290	305	325
Man-made fibres	Million kilogram	1244	1066	1268	1285
Raw wool	Million kilogram	45.20	45.00	45.00	45.00
Raw silk	Million kilogram	18.31	18.37	19.69	20.41
Production of yarn					
Cotton yarn	Million kilogram	2948	2896	3079	3490
Other spun yarn	Million kilogram	1055	1014	1114	1223
Man-made filament yarn	Million kilogram	1509	1418	1523	1550
Production of fabric					
Cotton	Million square meter	27,196	26,898	28,914	31,718
Blended	Million square meter	6888	6766	7767	8278
100 % non-cotton (including khadi, wool and silk)	Million square meter	21,173	21,302	23,652	21,765
Total	Million square meter	**55,257**	**54,966**	**60,333**	**61,761**
Per capita availability of cloth	Square meter	41.85	39.01	43.12	43.96
Production of textile machinery	Rs. crore	6155	4063	4245	6150
Textile exports					
Including jute, coir and handicraft	Million US$	22,147	20,942	22,099	26,561
Textile imports					
Including jute, coir and handicraft	Million US$	3327	3527	3432	4173

Source Third Handloom Census of India 2010, National Council of Applied Economic Research and Ministry of Textiles

Note Exports/Imports datasource DGCI&S, Kolkata—Foreign Trade Statistics of India (Principal Commodities and Countries)

[a]Cotton year

Table 2 State-wise distribution of textile industries in India

State-wise, category-wise and management-wise number of cotton/man-made fibre textile mills (non-SSI) as on 31/03/2011															
States/union territories/cities	Number of textile mills														
	Spinning					Composite					Total				
	Central	State	Co-op.	Private	Total	Central	State	Co-op.	Private	Total	Central	State	Co-op.	Private	Total
States															
Andhra Pradesh	2	0	5	143	150	0	0	0	2	2	2	0	5	145	152
Assam	0	3	1	1	5	0	1	1	0	2	0	4	2	1	7
Bihar	1	0	3	1	5	0	1	0	0	1	1	1	3	1	6
Chhattisgarh	0	0	0	1	1	0	0	0	0	0	0	0	0	1	1
Delhi	0	0	0	0	0	0	0	0	0	0	0	0	0	0	0
Goa	0	0	0	1	1	0	0	0	0	0	0	0	0	1	1
Gujarat	1	0	0	42	43	1	12	0	34	47	2	12	0	76	90
Haryana	0	0	0	69	69	0	0	0	2	2	0	0	0	71	71
Himachal Pradesh	0	0	0	17	17	0	0	0	0	0	0	0	0	17	17
Jammu and Kashmir	0	0	0	2	2	0	0	0	0	0	0	0	0	2	2
Jharkhand	0	0	0	1	1	0	0	0	0	0	0	0	0	1	1
Karnataka	1	1	12	33	47	2	0	0	5	7	3	1	12	38	54
Kerala	4	8	5	14	31	1	0	0	2	3	5	8	5	16	34
Madhya Pradesh	1	0	3	40	44	1	1	0	11	13	2	1	3	51	57
Maharashtra	5	1	83	64	153	9	5	0	23	37	14	6	83	87	190
Manipur	0	1	0	0	1	0	0	0	0	0	0	1	0	0	1
Orissa	1	2	6	6	15	0	1	0	0	1	1	3	6	6	16
Punjab	2	0	2	87	91	0	0	0	6	6	2	0	2	93	97
Rajasthan	3	0	3	45	51	0	1	0	10	11	3	1	3	55	62
Tamil Nadu	7	1	18	897	923	1	1	0	28	30	8	2	18	925	953

(continued)

Table 2 (continued)

State-wise, category-wise and management-wise number of cotton/man-made fibre textile mills (non-SSI) as on 31/03/2011

States/union territories/cities	Number of textile mills														
	Spinning					Composite					Total				
	Central	State	Co-op.	Private	Total	Central	State	Co-op.	Private	Total	Central	State	Co-op.	Private	Total
Uttar Pradesh	2	11	11	32	56	3	0	0	7	10	5	11	11	39	66
Uttaranchal	0	2	0	6	8	0	0	0	1	1	0	2	0	7	9
West Bengal	3	3	2	15	23	0	0	0	7	7	3	3	2	22	30
Union territories															
Dadra Nagar Haveli	0	0	0	10	10	0	0	0	2	2	0	0	0	12	12
Daman & Diu	0	0	0	1	1	0	0	0	0	0	0	0	0	1	1
Pondicherry	1	2	2	4	9	0	1	0	0	1	1	3	2	4	10
Grandtotal	**34**	**35**	**156**	**1532**	**1757**	**18**	**24**	**1**	**140**	**183**	**52**	**59**	**157**	**1672**	**1940**
Ahmedabad	1	0	0	17	18	1	8	0	22	31	2	8	0	39	49
Mumbai	2	0	0	4	6	5	1	0	8	14	7	1	0	12	20
Coimbatore	5	0	2	316	323	1	1	0	11	13	6	1	2	327	336
Kanpur	0	0	0	0	0	3	0	0	1	4	3	0	0	1	4

Source Statistics reported by Confederation of Indian Textile Industry with data obtained from the Annual Report for 2010–2011 from Textile Commissioner's Office, Mumbai, India

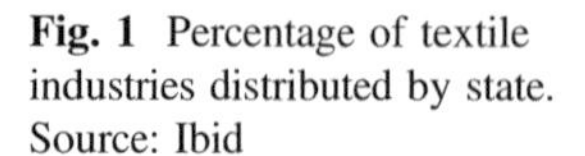
Fig. 1 Percentage of textile industries distributed by state. Source: Ibid

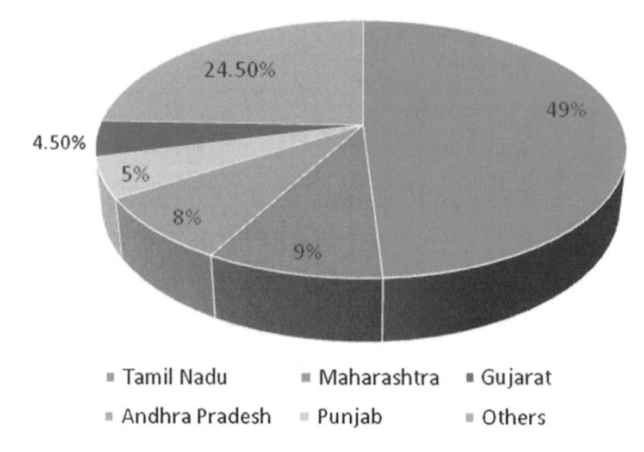

component of the textile industry is formed mainly by the decentralized power looms/hosiery and knitting sector.

The Indian textile industry is unique when compared to those in other countries because of the intimate relation it shares with the Indian agriculture sector in the way of its dependence for raw materials such as cotton. It also caters to the global demand for a wide range of textile products as required by various market segments, making the Indian textile industry truly versatile.

The fundamental function of the textile industry is to convert various fibres into yarn and further into fabrics and other products. This is followed by the dyeing and finishing processes which could be done either during the various phases of production or separately in the end. Therefore, the textile industry depends on a string of operation's to produce the final product.

The state-wise distribution of textile industries (Table 2) shows that Tamil Nadu, Andhra Pradesh, Maharashtra and Gujarat possess the maximum number of textile mills, accounting for more than 70 % of the total in the country as shown in Fig. 1.

It is during the processing stage of textile fibres that the operations use several chemicals such as dyes, auxiliary chemicals and sizing materials, to impart desired characters to fabrics. During the entire dyeing and printing process, large volumes of water are used for several processes such as wetting of fibres and dissolving dyes. This water is released in large amounts and causes most of the water pollution. The water released carries with it most of the toxic chemicals and therefore needs to be treated. The wastewater treatment is usually done through primary, secondary and tertiary stages. However, these treatment processes do not remove many of toxic materials such as dissolved solids and trace metals. This calls for advanced treatment technologies for the treatment of wastewater effluent from textile industries. Currently, studies are conducted to explore the possibilities of using nanomaterials for treating wastewater from textile industries. Some of the techniques include the use of nanochemicals for photocatalysis, nanosorbents for adsorption and zeolites for pollutants removal. Also, several treatment technologies are combined together to form the zero liquid discharge process to purify and recycle virtually all of the wastewater produced.

2 An Overview of the Textile Industry

The textile industry can be subdivided into different sectors based on the raw material being processed: the cotton industry, woollen industries, synthetic fabric industries, etc. These industries consist of a string of elaborate processes to produce the final product as shown in Fig. 2. The major process among them is the textile printing and dyeing process. Some of the steps involved in these two processes are pre-treatment, the dyeing and printing processes, finishing.

Pre-treatment is carried out by washing, desizing, scouring and other processes. These processes require large amounts of water. Dyeing is done to transfer the desired colour to the required fabric and produce the coloured fabric. For this purpose, the dye is dissolved in water. Printing on the other hand is a kind of dyeing

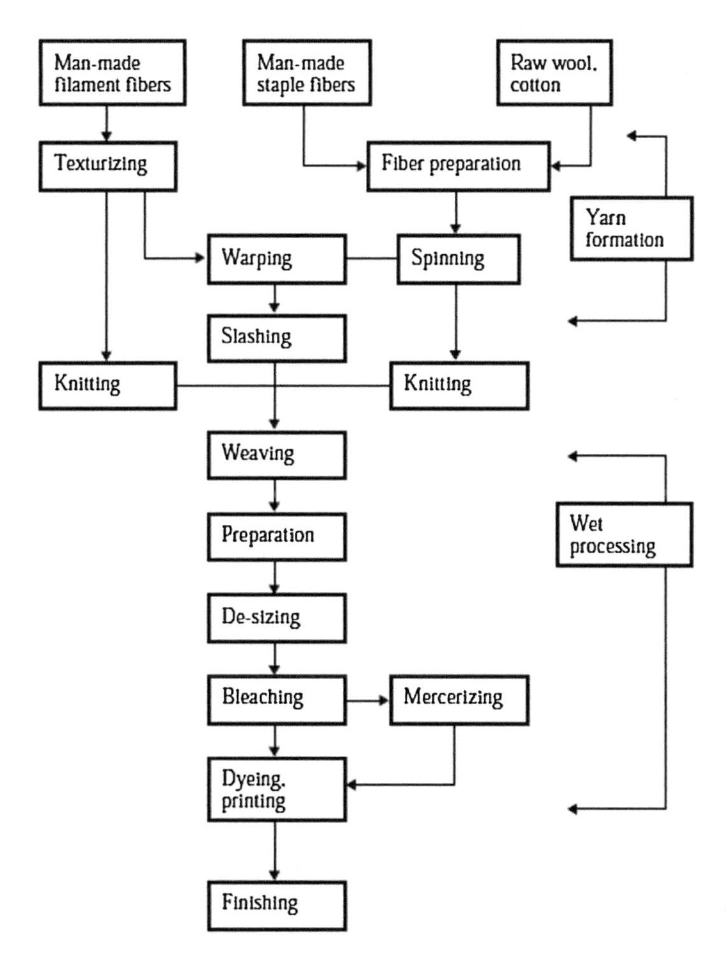

Fig. 2 Textile industry flow chart (Babu et al. 2007)

which is concentrated on a single portion of the fabric that makes up the design. In dyeing, the colour is applied using a solution of the dye, where the dye is suspended in water. Printing on the other hand employs the dye in the form of a thick paste. After the dyeing or printing processes, it is important for the fabric to undergo the finishing processes. The finishing processes are necessary to impart specific properties to the fabric such as strength, softness and durability.

The finishing process is carried out using several finishing agents for softening, creating cross-links within the fabric and waterproofing. All these processes require some amount of water and therefore can lead to release of harmful chemicals and toxins into the effluent stream, essentially requiring effective water treatment. In some cases, mercerization, base reduction processes are carried out before dyeing and printing.

Bleaching is the process of removal of any undesired colouring from fabric before dyeing or printing. Bleaching uses three major chemical procedures, namely sodium chlorite, sodium hypochlorite or hydrogen peroxide bleaching, with the first two procedures being the most common ones. These bleaching agents are carried in with water usually to either dilute or for better reaction with fabric due to the ease of adsorption on fabric. Chlorine dioxide acts as the oxidizing agent. However, chlorine dioxide being a powerful oxidizing agent is corrosive and highly harmful in nature. Therefore, proper disposal and treatment of this effluent are highly necessary.

3 Major Pollutants

The different processes in the textile industry lead to discharge of various toxic pollutants into the water stream which can be harmful. It is important to understand the nature of these pollutants in order to ensure the proper removal and application of the right treatment method (Tüfekci et al. 2007). The major pollutants are the organic chemicals such as azo dyes, pulp, gum, cellulose, hemicellulose and alkali. Water from dyeing and printing process accounts for more than 50 % of the total effluent (Saxena and Kaushik 2011).

Pre-treatment of polyester fibres is done by reduction reaction between the polyester fibre and 8 % sodium hydroxide at 90° for duration of 45 min. This leads to decomposition of the fibre into terephthalic acid and ethylene glycol. The COD of the resulting effluent accounts for about 60 % of the effluent from dyeing and printing processes even though it accounts for only 5 % by volume of the total effluent (Razzak 2014).

Chromium is another kind of pollutant in wastewater which causes a lot of concerns. Chromium is usually used as a catalyst or as chromium dyes for wool industries or comes from potassium dichromate used for tanning. Depending on the type of dye and process being employed, the chromium content is 200–500 times more in effluent than earlier due to the dyeing rate being ceased after process (Wang et al. 2011).

Table 3 Pollutants from various processes in textile industry (Wang et al. 2011)

Process	Compounds
Desizing	Sizes, enzymes, starch, waxes, ammonia
Scouring	Disinfectants and insecticides residues, NaOH, soaps, surfactants, fats, waxes, pectin's, oils, spent solvents, enzymes
Bleaching	H2O2, sodium silicate or organic stabilizer, high pH
Mercerizing	High pH, NaOH
Dyeing	Colour, metals, salts, surfactants, organic dyes, sulphide, acidity, alkalinity, formaldehyde
Printing	Urea, solvents, colour, metals
Finishing	Resins, waxes, chlorinated compounds, acetate, stearate, spent solvents, softeners

pH is another factor which needs to be considered in wastewater from dyeing processes. During the processes of scouring, desizing and mercerization, which are carried out before printing and dyeing, the pH of effluent wastewater stays around 10–11 when treated with alkali at 90°. As already mentioned, reduction of polyester fibres uses sodium hydroxide base which has a pH ranging between 10 and 11 (Menezes and Choudhari 2011). Therefore, the effluent wastewater is usually alkaline in nature and the first treatment process includes adjusting the pH value of the dyeing wastewater.

Nitrogen in ammonia and urea is also major pollutants and harmful. Batik and other complicated techniques use urea. The total nitrogen content in urea is 300 mg/L, which can be difficult to treat. Phosphorus is also present in wastewater which usually comes from the phosphor-containing detergents and monosodium phosphate used as buffer. Phosphorous can lead to rapid eutrophication of water bodies and therefore must be kept under check. Sulphide mainly comes from the sulphur used in sulphur dyes which are preferred due to their low price. However, sulphur is highly toxic and its use is forbidden in a number of developed nations. A list of the pollutants present in textile wastewater is given in Table 3.

3.1 *Colour/Dyes*

The presence of colour in the effluent from dyeing and printing processes is one of the main problems in textile industry. The colour of the dye comes from the chromophore in the dye. Colour in water is easily visible to the naked eye even at very low concentration. Hence, colour from textile wastes carries significant aesthetic importance. Most of the dyes are stable and is not degradable even under the effect of light or ozone. Another reason for concern is that the conventional treatment methods do not degrade the dyes completely. Due to this reason, removal of dyes from the wastewater effluent remains a major problem in most of textile industries. Aniline is one of the organic compounds which are mainly released from the dyes used. Dyes such as Congo red, amino or azo groups contain benzene rings which lead to increased carbon rings and nitrogen groups in the wastewater streams, which are extremely difficult to degrade.

3.2 Dissolved and Suspended Solids

Another critical parameter which has to be kept under constant check in effluent from textile industries is the amount of dissolved solids present. Common salt and Glauber salt, which are used for the recovery of dyes in textile industries, lead to a rapid increase in the total dissolved solids (TDS) of the effluent. However, just like dyes, it is difficult to treat TDS with conventional treatment systems. The presence of high TDS in the effluent can lead to an increase in TDS of surface water and groundwater sources which can be catastrophic as groundwater and surface water are subject to human consumption. The presence of dissolved solids in effluent may also be harmful to crops and in turn restrict the use of water for agricultural purposes.

The sources of suspended solids during the production process include undissolved raw materials, such as cellulose and pulp and fibre scrap. They can be removed by suitable mechanical separation methods. The secondary sedimentation tank outflow therefore contains large amount of suspended solids.

3.3 Toxic Metals

The effluent from textile industries contains metal ions as well. There are two main sources of metal ions. Firstly, the metals present in chemicals such as caustic soda, sodium carbonate and other salts which are used during alkali reduction or bleaching can be washed into the effluent as impurity. For instance, mercury can be present in caustic soda as impurity if it is produced using mercury cell processes (production of sodium hydroxide). Secondly, the source of metal could be from the dyes itself, i.e. the metalized mordent dyes. Most of the metal complex dyes contain a chromium base and as already discussed chromium can easily be washed into effluent from tanning industries.

3.4 Chlorine

The use of various chlorine carrying compounds such as sodium hypochlorite for bleaching in textile processing leads to the presence of residual chlorine in the effluent water stream. This chlorine-containing wastewater if released without treatment leads to reduction of dissolved oxygen from water bodies and reversely affects marine life. Chlorine in wastewater can also react with other substances in water to produce harmful and toxic materials.

3.5 Refractory Materials

Wastewater effluents from textile industries are often polluted with non-biodegradable organic materials which are also referred to as refractory materials. An example of refractory materials includes detergents. The presence of such refractory materials in turn increases the chemical oxygen demand (COD) of the effluent stream. Other organic contaminants, such as sizing materials, acids and enzymes are also present in the effluent. The level of these pollutants in wastewater is controlled by the use of suitable biological treatment processes.

4 Effluent Standards

4.1 Standards for Various Pollutants Present in Textile Effluent in India

Every country has a set of effluent standards for every industry, established by the pollution control board of the country. In India, the Central Pollution Control Board (CPCB), Ministry of Environment and Forests, Government of India, establishes the effluent standards for various industries. The effluent standards for effluent from cotton textile industries as per the Environment (Protection) Rules, 1986 set by the CPCB is given in Table 4.

The effluent hence released has to comply with these standards, and the treatment methods adopted have to ensure that they work around these standards. There are several specific standards depending on the final water body into which the effluent is discharged into. The trade effluent standards for being discharged into inland surface waters in the state of Tamil Nadu are given in Table 5.

Table 4 Effluent standards for various parameters in India

Common parameters	
pH	5.5–9
Suspended solids (mg/L)	100
Biochemical oxygen demand1[3 days at 27oC] (mg/L)	150
Oil and grease (mg/L)	10
Bioassay test (mg/L)	90 % survival of fish of after 96 h
Special parameters (mg/L)	
Total chromium as (Cr) (mg/L)	2
Sulphide (as S) (mg/L)	2
Phenolic compounds (as C_4H_2OH) (mg/L)	5

Table 5 Effluent standards for textile industries in Tamil Nadu, India

Parameter	Concentration not to exceed, milligram per litre (mg/l), except pH
pH	5.5–9.0
Total suspended solids	100
Biochemical oxygen demand (BOD)	30
Chemical oxygen demand (COD)	250
Total residual chlorine	1
Oil and grease	10
Total chromium as Cr	2
Sulphide as S	2
Phenolic compounds as C_6H_5OH	1

Source Standards for Discharge of Trade Effluents into Inland Surface Waters, Tamil Nadu Pollution Control Board (2015)

Table 6 Pollution standards for textile industry in Germany (Wang et al. 2011)

Serial number	Parameters	The limits of discharged concentration (mg/L)
1.	COD	160
2.	BOD	25
3.	TP	2.0
4.	TN	20
5.	NH_3	10
6.	Nitrite	1.0

4.2 Standards for Various Pollutants Present in Textile Effluent in Germany

The standards for various pollution parameters in textile industry effluent Germany is shown in Table 6.

4.3 Standards for Various Pollutants Present in Textile Effluent in the USA

The EPA publishes the effluent requirements using best practical control tech (BPT) in the USA, the details of which have been shown in Table 7.

Table 7 Discharge standards for textile industry in the USA (Wang et al. 2011)

Serial number	Parameters	Maximum load (kg/t of fabric)
1.	BOD_5	22.4
2.	COD	163
3.	TSS	35.2
4	S	0.28
5	Phenol	0.14
6.	pH	6.0–9.0

5 Conventional Treatment Methods

The wastewater effluent from textile dyeing and printing process contains large concentration of organic matter, most of which is difficult to degrade. The released effluent has high BOD and COD content which can be perilous to the state of aquatic life and further to human health. For instance, the presence of coloured water does not allow penetration of light and therefore restrains growth of life and the presence of phosphorus can accelerate Eutrophication. In order to check the adverse effects of these toxic chemicals in the discharged effluent, it is important to establish suitable treatment methods for textile wastewater.

The conventional treatment techniques that have existed in the past few decades include physicochemical treatment methods, biological treatment and tertiary treatment methods which employ a combination of methods.

5.1 Physicochemical/Primary Treatment

5.1.1 Screening

Treatment of wastewater involves a series of unit processes some of which are physical, chemical and biological in nature. An industry releases different streams from different points at different times, and hence, the first step in treatment of wastewater is to mix and equalize these streams. Some industries use screening process and oil trap processes before mixing various streams. Screening involves the use of meshes and sieves of appropriate size to hold back the grit and the undissolved large particulate matter or debris. Oil trap follows the same function but with the purpose of adhesion separation and removal of oil and grease.

5.1.2 Equalization

Equalization is performed to ensure the uniform distribution of pH, pollution load and temperature. It is carried out by air agitation or mechanical mixing. It helps to fix a single separation process as conditions remain a constant. The hydraulic retention time for this is around 8 h.

5.1.3 Floatation

Equalization is followed by the floatation process. In this process three phases are established of liquid, gas and solid due to the production of microbubbles. Compressed air is usually passed in through the mixture which causes smaller particles to attach to it due to surface tension, buoyancy of the rising bubble hydrostatic pressure. Due to lower density compared to water, these bubbles rise, helping to get rid of tiny fibres and oil particles.

5.1.4 Coagulation, Flocculation and Settling

After all these processes are complete, coagulating chemicals such as alum, ferrous sulphate, ferric chloride and lime are added to the system in a process called flash mixing. The colloidal particles which were not separated during settling and floatation are separated during this process. Colloidal particles usually carry a charge, either positive or negative and the addition of a coagulating agent exposes the charged colloid to an oppositely charged agent which leads to coagulation and further settling due to increase in density. Formation of flocs (aggregates) increases the density further and helps in rapid settling.

This mixture is then processed through the flocculator and further the coagulated mass is separated by settling in settling tanks. A clariflocculator helps to remove sludge content by some amount of adsorption. Choosing the right coagulant depends on the kind of pollutants present in the effluent stream after equalization. Coagulation or flocculation helps in removal of suspended solids and to a certain extent helps in removal of colour from the exit steam as well. They also help reduce the BOD and COD (Verma et al. 2012).

5.2 Biological Treatment

Biological treatment helps remove dissolved matter and is more efficient than physicochemical treatment. The ratio between organic load and the biomass present in the reaction tank, the reactor temperature and oxygen concentration determines the removal efficiency. Aeration leads to a suspension effect, but it is important to ensure that a mixing energy that can destroy the flocks is not reached, because it can inhibit the settling.

The biomass concentration usually is between a range of 2500–4500 mg/l and oxygen content is found to be around 2 mg/l. Based on the oxygen availability, biological treatment can be classified into two main types: aerobic and anaerobic treatment methods. Aerobic biological treatment has higher efficiency and wider application compared to anaerobic treatment methods and therefore is more widely used (Rai et al. 2005).

5.2.1 Aerobic Biological Treatment

There are different kinds of bacteria which can be classified based on their oxygen requirements. These include aerobic, anaerobic and facultative bacteria. As the name suggests, aerobic biological treatment employs aerobic bacteria and some facultative bacteria. Aerobic biological treatment is usually of two types: activated sludge process and biofilm process.

Activated Sludge Process

Activated sludge is composed of an aggregate mixture of a number of micro-organisms and has the ability to decompose and adsorb organic matter (Pala and Tokat 2002). Once the activated sludge is formed, it can be removed and the remaining water further purified. The oxidation ditch process and sequencing batch reactor (SBR) process are types of activated sludge processes.

Oxidation ditch process
Oxidation ditch process is a type of activated sludge process which was first developed in the Netherlands. The components of an oxidation ditch process include an aeration equipment, an equipment for the water which has to be purified to come in or go out and mixing equipment. The ditch can be of different shapes which include a ring structure, L shape, rectangular or round shapes. Wastewater activated sludge and various microbes are mixed in a continuous loop ditch. This completes nitrification and denitrification of wastewater. The oxidation ditch resembles a plug flow, completely mixed oxidation tank in nature.

The advantages of this process include long hydraulic retention time (HRT), long sludge age and low organic loading which in turn lead to high degree of purification, high reliability, easier maintenance, high impact resistance, stability, lower investment, easy operation and management, and lower energy consumption. Due to the formation of discrete aerobic, anoxic and anaerobic zones, the oxidation ditch process performs denitrification efficiently.

Sequencing Batch Reactor Process
SBR treatment process leads to removal of COD and colour from effluent water. This process has good shock loading which means it can withstand good amount of organic matter and reserve water. Also, the total residence time, run time and gas supply of each stage can be altered based on the quantity of water coming in or going out. The original water contains large amount of organic matter, which provides good condition for the bacteria to grow. This activated sludge leads to low sludge production compared to the standby phase when the sludge is in the endogenous respiration

phase, rendering low sludge yield. Other advantages of this process include less processing equipment, simple structure requirement and ease of operation.

Biofilm

In this process various micro-organisms are made to attach to a single support and the wastewater is purified when it flows on this surface by contact. The biofilm processes can further be divided into three categories which include biological fluidized beds, biological contact oxidation and rotating biological contactors.

Biological Fluidized Bed

The fluidized bed process is also referred to as the suspended carrier biofilm process (Chaohai et al. 1998). This is a new and highly efficient wastewater treatment process which uses the modern fluidization technology whose principle is to keep the whole system at a fluidized state and in turn enhance the contact time and area between solid and fluid particles. The substances usually used as carrier or filling in the system include carbonaceous materials such as sand, activated carbon, anthracite or other particles. The diameter of these particles needs to be less than 1 mm.

A pulse of the wastewater is allowed to pass through the carrier which leads to fluidization of the carrier particles and in turn leads to formation of biofilm on these particles. Smaller size of carrier particles is preferred as it leads to higher surface area (the preferred area is between 2000 and 3000 m^2/m^3), which in turn can hold higher amount of biomass. This process yields efficiency of 20–30 times higher than the conventional treatment processes and hence can be used for treating high consistency organic effluent from textile industries.

Biological Contact Oxidation

Biological contact oxidation is a process which is primarily used for treating wastewater from the dyeing process (Guosheng 2000). Fillers are set in tanks called aeration tanks and the set-up mimics that of the activated sludge process. A biofilm is formed on the fillers placed in the tank and the wastewater in the tank contains a certain quantity of activated sludge. Aerobic micro-organisms are present in the biofilm and these microbes degrade the organic matter present in wastewater when they come in contact with each other.

Rotating Biological Contactor

Rotating biological contactor is a wastewater treatment process which is modelled based on the conventional biological filter (Abraham et al. 2003). It consists of a series of discs which are usually composed of low weight materials such as plastic or glass platers. The diameter of the discs used ranges from 1 to 3 m. The disc is constructed such that half of the disc area is exposed to air while the other half is submerged in the supplied wastewater. The Rotating device makes the discs rotate slowly in the horizontal axis which in turn helps to completely mix the wastewater due to continuous rotational motion. The micro-organisms present in the wastewater form a biofilm on the disc when the disc is exposed to the wastewater. When the same disc is exposed to air, the biofilm comes in contact with oxygen and gets adsorbed. The micro-organisms oxidize and degrade the organic matter present in wastewater by catalysis reaction using enzymes and release the metabolite at the same time. Hence, an adsorption–oxidation–oxidative decomposition process which is followed continuously is used to treat the wastewater.

The rotating biological contactor has good power saving, has high shock load capability, produces very little sludge, is easy to maintain and manage and produces very little noise. The shortcomings of this equipment include the need for larger-area and well-maintained buildings. Main parameters which affect the performance of the RBC are speed of rotation, wastewater residence time, area of disc submerged and temperature.

5.2.2 Anaerobic Biological Treatment

The anaerobic biological treatment process is carried out in the absence of oxygen. Bacteria called anaerobic bacteria are used to degrade organic matter under anaerobic conditions. This process has been used for treatment of both high and low concentration organic wastewater such as textile dyeing wastewater (Delée et al. 1998). Effluent from textile industries contains organic matter with organic matter content as high as 1000 mg/L. Anaerobic biological treatment is usually adopted for the treatment of high concentration wastewater while aerobic treatment is used to treat wastewater with low organic matter.

Presently the most widely used technique for anaerobic biological treatment is the hydrolysis acidification process. In this process, the anaerobic and facultative bacteria decompose the biodegradable organic matter present in wastewater into small molecular organic matter. This principle can be used for biodegradation of coloured groups of dye molecules present in textile dyeing wastewater. The acids released by these bacteria can also be used to reduce the amount of alkalinity in textile effluent and in turn make the water more suitable for marine life. Other processes used for treatment of effluent from textile industries include upflow anaerobic fluidized bed (UABF), upflow anaerobic sludge bed reactor (UASB) and anaerobic biological filter.

6 Advanced Treatment Methods

6.1 Adsorption

Adsorption is a surface phenomenon which is used to remove suitable organic matter from wastewater effluents (Kumar et al. 2014; Oliveira et al. 2002). Chemicals such as cyanides and phenol derivatives can be removed using this process. This process is particularly useful as it can be used to remove compounds difficult to remove by conventional treatment methods. As the wastewater is passed through the adsorbent, the organic pollutants keep getting adsorbed. Activated carbon is the most widely used adsorbent. Activated carbon is usually produced by treatment of petroleum products or wood. The material is burned in the presence of air, creating a char. Oxidation of this charred substance at high temperature gives a porous substance which is activated carbon. An important characteristic of activated carbon is its high surface area which is important for adsorption of the organic matter onto the adsorbent.

The activated carbon once completely used up, i.e. when all its pores are clogged requires regeneration. Regeneration is usually done chemically or thermally; however, latest methods include solvent, electrochemical, microwave induced and supercritical regeneration. Suitable acids or chemicals can be passed to partially regenerate the adsorbent and this process leads to need for frequent regeneration (Leng and Pinto 1996). In case of thermal regeneration, the spent adsorbent (activated carbon) is transported in water slurry to a regeneration unit. In this unit, the adsorbent is dewatered by application of heat using a furnace and reheated under controlled conditions (Dabrowski et al. 2004). This complete process oxidizes the impurities and allows the volatile substances to escape leaving adsorbent with increased pore volume. In order to remove excess heat, the hot adsorbent is quenched with water and moved back to the column. This process has higher efficiency than chemical regeneration. Other materials used as possible adsorbents include clay, silica and fly ash.

6.2 Ion Exchange

The ion exchange process is another modern method of removal of inorganic salts, heavy metals and few specific organic anionic compounds from wastewater (Dabrowski et al. 2004). Salts are usually ionic compounds usually composed of a positive ion from an alkali and a negative ion from an acid. The ion exchange resins or membranes have the ability to exchange suitable ions present in electrolytic solutions, in this case wastewater with soluble ions. A cation exchanger when contacted with an electrolyte such as calcium chloride will scavenge the positively charged ion, in this case calcium, from the electrolyte and in turn replace it with the soluble ions which in this case is sodium. Hardness in water arises from salts of calcium and magnesium, and this method provides a convenient technique to reduce hardness in wastewater. Ion exchange resins can be made of different compounds such as phenolic compounds and sulphonic styrene's.

6.3 Membrane Filtration

6.3.1 Reverse Osmosis

Reverse osmosis follows the principle that certain semipermeable membranes are selective in nature and allow the flow of only selective ions through them

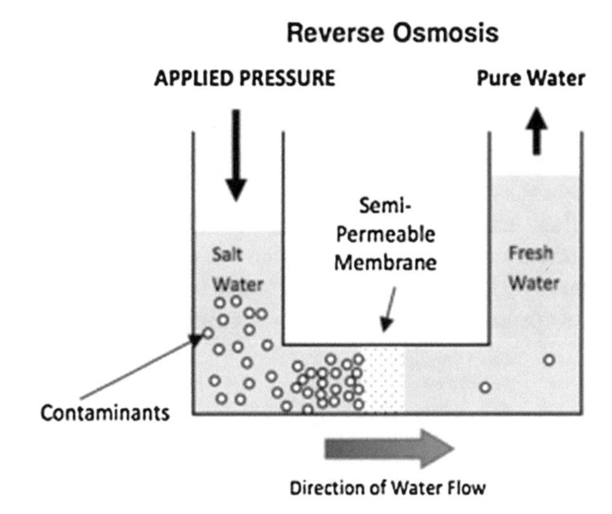

Fig. 3 Reverse osmosis (Wayman 2015)

(Fritzmann et al. 2007). The commonly used membranes include nylon and cellulose membranes such as cellulose acetate. The wastewater to be treated is passed at high pressure through the semipermeable membrane as shown in Fig. 3.

The applied pressure has to be high enough to act as a driving force to overcome the overall osmotic pressure of the stream. When this criterion is met, water flows from the wastewater chamber to the clear water chamber through the semipermeable membrane. A reverse osmosis system consists of a pre-treatment section through which the wastewater is passed to remove suspended solids by mechanical separation methods and if necessary, ions such as iron and magnesium. This is necessary to ensure that these suspended solids and ions do not foul the semipermeable membrane. After pre-treatment the wastewater is pressurized and sent to the reverse osmosis chamber. When the applied pressure is greater than the osmotic pressure across the membrane, clear water passes from wastewater to the clear water chamber, emerging at atmospheric pressure.

Reverse osmosis can be used to treat effluent from primary, secondary and tertiary stages of wastewater treatment. The reverse osmosis models include spiral wound system, tubular systems, Hollow fibre membranes and the disc module. The main downfall of this process is fouling of the membranes in use due to organic materials, colloids and micro-organisms. Also scale formation can take place due to hardness-causing constituents (carbonates) in the effluent stream. Chlorine and ozone can oxidize the membrane and hence should not be present in wastewater to be treated.

6.3.2 Ultrafiltration

Ultrafiltration and reverse osmosis follow similar techniques. The difference between the two processes is primarily the retention properties of the membranes.

Ultrafiltration membranes have higher pore size and can retain only macro-molecules and suspended solids. Reverse osmosis membranes on the other hand have smaller pore size and hence can be used for removal of all solutes including salts. The osmotic pressure difference across an ultrafiltration membrane is negligible as the salts pass through along with the permeate water. This makes it necessary for a much lesser pressure to be applied for the process to take place (Babu et al. 2007). Ultrafiltration membranes materials include cellulose derivatives such as cellulose acetate, polymers of nylon and other inert polymers. This makes it possible to treat acidic or caustic streams and not limiting the process unlike that for reverse osmosis due to chemical attack of the membranes.

6.3.3 Nanofiltration

When the salt concentration in the permeate water is not of prime importance as compared to the reverse osmosis operation, nanofiltration can be used for wastewater treatment (Nguyen et al. 2012). Hardness-causing elements in water are salts of calcium and magnesium, and nanofiltration can be used to remove these salts. It can also be used for the removal of micro-organisms such as bacteria and viruses. Nanofiltration is suitable to use when permeate with TDS, but without colour, COD and hardness is acceptable. It is a more cost-effective method. For higher efficiency, turbidity and the amount of colloidal content in wastewater should be low.

Fouling is a major problem whenever membrane separation techniques are used for purifying wastewater (Jiraratananon et al. 2000). Wastewater has dissolved elements such as silica, calcium, barium, iron and strontium which may precipitate on the surface of the membrane and in turn clog the pores of the membrane. This phenomenon is termed as fouling of the membrane. Bacteria present in wastewater which proliferates in warm environmental conditions can also cause fouling of membranes. This may cause increased pressure difference across the membranes and reduced flux leading to higher expenditure. Fouling can be avoided or reduced by selecting a suitable chemical dose to counteract the deposition and precipitation of organic matter. Antiscalents can be used for the removal of mineral scales. Back washing with acid also reduces fouling.

6.4 Ozonation

Ozone is known to be one of the strongest oxidizing agents commercially available and it can be used for disinfection purposes on wastewater. Ozonation can be used to break down complex large molecules such as phenols (Langlais et al. 1991). Oxidation of organic and inorganic materials, removal of odour and removal of colour are the main applications of ozone in industries. Naturally, ozone is present as an unstable gas; under normal conditions, it decomposes to give molecular

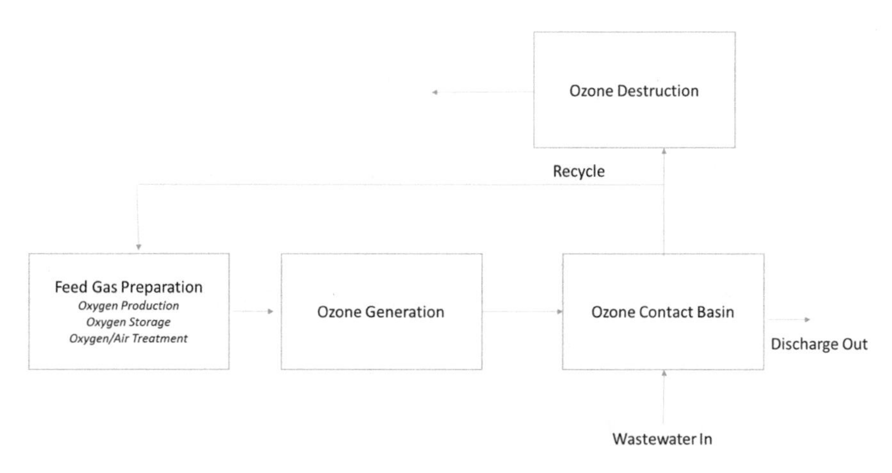

Fig. 4 Treatment using ozonation

oxygen and nascent oxygen. Due to this property, ozone has to be produced on the site for industrial applications. There are several methods of producing ozone, but the corona discharge method is the most widely used. An ozone generation unit usually consists of a series of electrodes. These electrodes are fit with cooling arrangements mounted in a gas tight container. The source gas use to produce ozone is usually air or oxygen. When this gas is passed between two electrodes which are separated by a narrow gap between them, the oxygen gets converted into ozone. The ozone released by the ozone generation unit is brought in contact with wastewater in the ozone contact basin by means of diffuser tubes or turbine mixers as shown in Fig. 4. Complete removal of colour and pollutants present in wastewater can be achieved by using 2 mg/l of ozone. The treated water is treated with sand (sand filtration) for the final cleaning.

6.5 Evaporation

6.5.1 Multiple-Effect Evaporation

In evaporation, the wastewater is heated in multiple-effect evaporators with the concentrate from one evaporator being fed to the next evaporator while the clear water is allowed to exit the system for further treatment. The fresh liquor and steam are added only to the first evaporator. The steam obtained by evaporation from the first stage acts as input steam for the next stage. The vapours and used steam are collected in the condenser.

Evaporators reduce the volume of the effluent and leads to an increase in concentration of salts in the stream. This makes their recovery more viable whenever intended. One problem of evaporation is the formation of scales on the walls and tubes of the evaporator due to the presence of salts of calcium and magnesium is wastewater. This leads to a reduction in rate of heat transfer and in turn the evaporation rate is also reduced. Cleaning is required for removal of these scales. Cleaning can be of both chemical and mechanical type. Chemical means involve acid treatment where acid is poured along the scale to loosen it from the surface of the tube. A scale cutter is a device mounted on a flexible shaft which is inserted from the top of the pipe. This device is moved up and down repeatedly to remove the scale in the entire length of tube. Once all the stages have been cleaned, the evaporator is closed and tested to check for any leaks. Proper seals are made to close these leaks.

6.5.2 Mechanical Vapour Compression

The mechanical vapour compression technique is used to concentrate the wastewater effluent. A heat exchanger is used which also acts as an evaporator or condenser. The heat is provided by the simultaneous condensation of the distillate. Latent heat is exchanged between hot and cool water in the evaporation–condensation process. A compressor which is present provides the energy required for separating the solution by heat transfer.

6.6 Crystallization

In crystallization, solid crystals are obtained from a homogeneous solution. It is a solid–liquid separation technique. Crystallization occurs only when a solution is supersaturated. Supersaturation of a solution is its state in which the solvent has higher concentration of solute than what can be dissolved in it at that temperature. By using this method, salts can be recovered from the mother liquor. Crystallizers are of two types, namely single-stage and multi-stage. The common chemicals removed from the effluent by crystallization include calcium sulphate, sodium chloride, calcium chloride and sodium sulphate. Therefore, the scale causing chemicals can be removed from wastewater and this can be used a pre-treatment method for wastewater being used for membrane purification and evaporation.

6.7 Zero Liquid Discharge (ZLD)

Zero liquid discharge is a modern separation technique which leads to a solid (or near solid) waste stream and purified water stream which can be treated further or used for domestic applications (Dalan 2000). Reverse osmosis and thermal methods can be used for concentrating the waste stream and formation of pure water. However, since reverse osmosis is more cost-effective, it is more widely used

(Heijman et al. 2009). ZLD consists of a series of wastewater treatment techniques which include microfiltration (MF) for pre-treatment, reverse osmosis (RO) for pre-concentration of waste solutions followed by the application of a multiple-effect evaporator. The major objective is to produce a solid concentrated stream which can be sent for land-based disposal and a relatively clean water stream for domestic use or sewer discharge. This leaves a clean effluent stream without any dissolved solids for further treatment while the filtered or concentrated solids can be sent for suitable disposal.

7 Application of Nanochemicals for Effluent Treatment

Nanoscience is the field of science which deals with the study of materials at the nanoscale. The smaller size of nanomaterials highly influences their properties as compared to their bigger counterparts. Nanotechnology is an application of nanoscience and it deals with the control, integration and manipulation of various atoms and molecules to form different materials at the nanoscale (Hornyak et al. 2008, 2009). The past decade has been subjected to drastic increase in the application of nanotechnology in applied science, especially in the area of water purification, opening up a new potential alternative to treat wastewater more efficiently and cost-effectively (Baruah et al. 2012; Baruah and Dutta 2016) Some of the key advantages of nanochemicals include their small size, high reactivity, high accuracy and most importantly, their capability to be produced by environment-friendly techniques, most of which are potentially cost-effective as well. Two of the promising applications of nanochemicals for wastewater treatment in textile industries are:

- Photocatalysis
- Nanosorbents.

7.1 Photocatalysis

Photocatalysis is a process which can be defined as a "change in the rate of a chemical reaction or its initiation under the action of ultraviolet, visible, or infrared radiation in the presence of a substance—the photo catalyst—that absorbs the light and is involved in the chemical transformation of the reaction partners" (McNaught and Wilkinson 1997). It is a process which is used to degrade pollutants in wastewater. A nanostructured catalyst medium which is sensitive to exposure to light is mainly employed.

The catalyst medium usually in use is a semiconductor material, which generates an electron–hole pair upon absorption of light with energy higher than its band gap energy. The electron–hole pair then reacts with water to produce highly reactive oxidizing and reducing radicals, such as super oxides, hydroxyl ions or other radicals. These radicals then degrade any organic/inorganic pollutant molecules present in the contaminated water through some secondary reactions. Photocatalysis is a surface phenomenon and its general mechanism is a complex process, which involves five basic steps (Pirkanniemi and Sillanpää 2002).

- Reactants diffusing to the surface of the catalyst
- Reactants getting adsorbed on the surface of the catalyst
- Reaction taking place at the surface of the catalyst
- Products undergoing desorption from the surface of the catalyst
- Products diffusing away from the surface of the catalyst.

Nanostructured semiconductor materials possess higher surface area which enables them to have better photocatalytic properties compared to their larger bulk counterparts. This property makes it possible for the generated electrons and holes to be available at the surface of the nanophotocatalyst rather than to be available in the bulk of the material. An ideal photocatalyst should exhibit the following properties:

- high photoactivity
- biological and chemical inertness
- photostability
- non-toxicity
- cost-effectiveness.

Some examples of typically used nanostructured semiconductor photocatalysts are zinc oxide (ZnO), titanium dioxide (TiO_2), zinc sulphide (ZnS), ferric oxide (Fe_2O_3) and cadmium sulphide (CdS).

The wide band gap semiconductors absorb light in the UV region of the solar spectrum. However, using high-energy UV light sources to excite the photocatalysts may not be a cost-effective solution in all cases. Therefore, research is currently focusing on using visible portion of the solar spectrum for conducting photocatalysis experiments. The advantages of using solar light for photocatalysis are that solar energy is free and abundantly available. Some of the key areas in which photocatalysis can be used for the treatment of textile wastewater include removal of:

- Organic contaminants.
- Inorganic contaminants.
- Heavy metals.

7.1.1 Removal of Organic Contaminants

Photocatalysis is widely used for the decomposition of harmful organic contaminants present in wastewater into harmless by-products, mostly carbon dioxide and water. The organic contaminants that can be removed from wastewater using photocatalysis include carboxylic acids, alcohols, phenolic derivatives and chlorinated aromatic compounds (Bhatkhande et al. 2002; Mills et al. 1993). Dyes being the major pollutants released into the effluent stream have to be treated and the nanophotocatalysts shown to be successful in this regard are semiconductor metal oxides such as TiO_2 and ZnO.

Additionally, photocatalysis has been employed to degrade natural organic matter also called as humic substances. Humic substances are naturally occurring yellow–brown organic materials with high molecular weight. TiO_2 nanoparticles have made it possible to achieve almost 50 % reduction in humic acid concentration in drinking water, observed under irradiation from a mercury lamp (Eggins et al. 1997).

7.1.2 Removal of Inorganic Contaminants

Inorganic contaminants include chemicals such as halide ions, cyanide, thiocyanate, ammonia, nitrates and nitrites that can be effectively decomposed by photocatalytic reaction (Hoffmann et al. 1995; Mills et al. 1996). The photocatalytic removal of toxic Hg(II) and CH3Hg(II) chlorides using TiO_2 nanoparticles under simulated solar light (AM1.5) (Serpone et al. 1987) and the removal of toxic potassium cyanide and Cr(VI) ions from water using visible light and ZnO nanoparticles have been achieved. Recently it has been found that CdS/titanate nanotubes are useful in photocatalytic oxidation of ammonia in water (Lee et al. 2002).

7.1.3 Removal of Heavy Metals

Heavy metals present in wastewater streams is another area of concern as it directly affects human health and presents a challenge for treatment plants, since the amount of can vary, depending upon the type of industry. However, due to the rare availability and high cost of some metals, recovery of the metals is mostly preferred over removal of the metals itself. Photocatalysis has been shown to successfully remove heavy metals in several cases. TiO_2 catalyst dispersion has been shown to help in the recovery of gold(III), platinum(IV), and rhodium(III), from a mixture of gold(III), platinum(IV), and rhodium(III) chloride salts (Borgarello et al. 1986). Recovery of gold from samples containing cyanide ions was also performed along with the degradation of CN^- by employing two peroxides, H_2O_2 and S_2O_2. TiO_2 has also been used for removing cadmium (Cd) from effluents by irradiating the stream with a source producing light rays of 253.7 nm wavelength. A combination of activated carbon obtained from sewage sludge along with TiO_2 nanoparticles has been used to reduce Hg^{2+} ions (Asenjo et al. 2011) followed by recovery of metallic Hg(0).

7.2 Nanosorbents

Sorption is a process in which a material referred to as sorbate gets adsorbed on another substance, called sorbent, by some physical or chemical interaction. Sorbents are used extensively for the removal of organic pollutants from wastewater such as those from textile industries. The sorption process usually takes place in three steps:

- Transport of the pollutant particle from water to the surface of the sorbent
- Adsorption of pollutant at the surface of the sorbent
- Transport within the sorbent.

Nanoparticles possess higher specific surface areas than the bulk particles. They also have the ability to functionalize easily with a variety of chemical groups and therefore improve their affinity towards the target contaminants. The aforementioned two properties make them very effective as sorbents. Moreover, nanosorbents possess nanosized pores that aid the sorption of contaminants. Nanosorbents have the ability to be used again through a regeneration operation where the absorbed pollutants are removed by suitable processes. Some of the nanosorbents are listed below.

7.2.1 Carbon Nanosorbents

Carbon-based nanomaterials are extensively used for the adsorption of various organic and inorganic pollutants in wastewater. Out of the carbon-based nanomaterials used, activated carbon is very popular because of its high adsorption capacity, high thermal stability, excellent resistance against attrition losses and low cost. Benzene and toluene are used as solvents for mixing dyes in the textile industry. Removal of benzene and toluene from the effluent stream is important as they add to the organic matter content and can be harmful. Adsorption of benzene and toluene from industrial wastewater on activated carbon has been carried out (Kadirvelu et al. 2001) and high adsorption capacity for benzene ($\sim$400–500 mg/g) and toluene ($\sim$700 mg/g) was noticed. Activated carbon has been proven effective for the removal of heavy metal ions, such as Hg(II), Ni(II), Co(II), Cd(II), Cu(II), Pb(II), Cr(III) and Cr(VI) (Kobya et al. 2005; Al-Omair and El-Sharkawy 2007; Zhong et al. 2006).

7.2.2 Metal Oxide Nanosorbents

The common metal oxides used as adsorbents are mostly oxides of iron (Fe), manganese (Mn), silicon (Si), titanium (Ti) and tungsten (W). As adsorbent materials, metal oxides have the advantages of being low-cost materials and can be functionalized easily to tune their adsorption capacity and selectivity. Fe oxides nanosorbents have recently been tested for the removal of several organic pollutants in wastewater (Luo et al. 2011; Zhang et al. 2010, 2013; Iram et al. 2010). The magnetic properties of Fe oxide nanosorbents allow them to be magnetically separated from water (Hu et al. 2010) The Fe oxides also showed excellent adsorption capacity for heavy metal ions (Nassar 2010; Jeon and Yong 2010). Nanostructured tungsten oxide (WO_2) has also displayed very high adsorption capacity for organic dyes in water (Abdolmohammad-Zadeh et al. 2013). A zinc–aluminium layered double hydroxide nanosorbent has been developed and successfully applied for the removal of reactive yellow dyes from several textile wastewater streams (El-Safty et al. 2012). Applications of non-metallic oxide, such as silica (SiO_2), as nanosorbents have also shown promising results in removing organic pollutants and heavy metals present in effluents (Yantasee et al. 2010; Zamzow et al. 1990; Maliou et al. 1992; Ouki and Kavannagh 1997; Ibrahim and Khoury 2002).

7.2.3 Zeolites as Sorbents

Zeolites have high specific surface area and high ion exchange capacity, making zeolites a sought after adsorbent for water treatment. Most of the zeolites occur naturally and can also be produced commercially. Zeolites are used widely for the adsorption of heavy metal ions (Pansini et al. 1991). The adsorption of lead and cadmium using two natural zeolites chabazite and clinoptilolite has been studied (Kesraoui-Ouki et al. 1993). Using the two natural zeolites pre-treated with NaOH, the authors demonstrated very high adsorption capacity for lead (Pb) and cadmium (Cd), with metal removal efficiency of more than 99 %. The high porosity of zeolite gives it a higher adsorption capacity, and the photocatalytic reduction ability of zeolite aids in reducing higher valence metal ions to the corresponding lower ones, thus decreasing their toxicity.

8 Nanochemicals Used for Textile Wastewater Treatment

Several commercial and advanced technological developments are employed for water treatment; however, nanotechnology has been established as one of the most advanced wastewater treatment techniques. Developments in nanoscale research has paved way for economically feasible and environmentally stable treatment technologies for effectively treating wastewater meeting the ever increasing water quality standards. Nanotechnology can possible address many of the water quality issues by using different types of nanoparticles. Nanotechnology uses materials of sizes smaller than 100nm in at least one dimension meaning at the level of atoms and molecules as compared with other disciplines such as chemistry and materials science (Masciangioli and Zhang 2003; Eijkel and Van den Berg 2005).

At this scale, materials possess new and significantly changed physical, chemical and biological properties mainly due to their structural variation, higher surface area-to-volume ratio offering several uses for pollution control such as treatment, remediation and detection (Rickerby and Morrison 2007; Vaseashta et al. 2007). These unique properties of nanomaterials, for example, high reactivity and strong sorption, are explored for application in water/wastewater treatment based on their functions in unit.

Nanoparticles can penetrate deeper and thus can treat water/wastewater which is generally not possible by conventional technologies. Their higher surface area-to-volume ratio enhances the reactivity with environmental contaminants. Nanotechnology has the potential to provide both water quality and quantity in the long run through the use of, for example, membranes enabling water reuse, desalination. In addition, it yields low-cost and real-time measurements through the development of continuous monitoring devices (Riu et al. 2006; Theron et al. 2008). Nanoparticles, having high absorption, interaction and reaction capabilities, can behave as colloid by mixing mixed with aqueous suspensions and they can also display quantum size effects (Alivisatos 1996). Energy conservation leading to cost savings is possible due to their small sizes; however, overall usage cost of the technology should be compared with other techniques in the market (Crane and Scott 2012).

Membrane technology, considered as one of the advanced water/wastewater treatment processes due to its efficient and low-cost filtration technique (Allabashi et al. 2007), has been developed to be even more efficient using nanomaterials. Nanoparticles have been frequently used in the manufacturing of membranes, allowing permeability control and fouling resistance in various structures and relevant functionalities (Li et al. 2009; Kim et al. 2008). Both polymeric and inorganic membranes are manufactured by either assembling nanoparticles into porous membranes or blending process. The examples of nanomaterials used in this formation include, for example, metal oxide nanoparticles such as TiO_2. CNTs have resulted in desired outputs of improved permeability, inactivation of bacteria and so forth (Chae et al. 2009; Barhate and Ramakrishna 2007).

Finally, nanofibrous media have also been used to improve the filtration systems because of the high permeability and small pore size properties they possess (Escobar et al. 2001). They are synthesized by a novel fabrication technique called electrospinning. Depending on the polymers selected, they exhibit different properties. In short, the development of different nanomaterials such as nanosorbents, nanocatalysts, zeolites, dendrimers and nanostructured catalytic membranes has made it possible to disinfect disease-causing microbes, removing toxic metals, and organic and inorganic solutes from water/wastewater. An attempt is made to highlight the factors that may influence the efficiency of the removal processes based on the available literature in the following section.

8.1 Disinfection

Biological contaminants are generally classified under three categories, namely micro-organisms, natural organic matter (NOM) and biological toxins. Microbial contaminants include human pathogens and free living microbes (Majsterek et al. 2004; Berry et al. 2006; Dugan and Williams 2006; Srinivasan et al. 2008; Jain and Pradeep 2005). Adsorbents such as activated carbon have reasonably good removal efficiencies and again a number of factors influence the removal process.

Groundwater can get easily polluted by bacteria and protozoa. The easiest method for treating micro-organisms is using oxidizing agents such as chlorine and ozone. However, chlorine dioxide is expensive and results in the production of hazardous substances such as chlorite and chlorate in manufacturing process. Ozone, on the other hand, has no residual effects but produces unknown organic reaction products. For UV disinfection, longer exposure time is required for effectiveness and also there is no residual effect. So, advanced disinfection technologies must, at least, eliminate the emerging pathogens, in addition to their suitability for large-scale adoption. There are many different types of nanomaterials such as Ag, titanium and zinc capable of disinfecting waterborne disease-causing microbes present in effluent. Their antibacterial properties arise from their charge. TiO_2 photocatalysts and metallic and metal oxide nanoparticles are a set of promising nanomaterials that exhibit antimicrobial properties. The ability of metal ions to produce the desired results in water disinfection has been highlighted by many researchers (Sondi and Salopek-Sondi 2004). The possible nanomaterials that can be utilized as disinfectants are listed.

8.1.1 Silver Nanoparticles

Silver has low toxicity and microbial inactivation and hence is the most widely used nanomaterial. Silver nanoparticles can be derived from its salts such as silver nitrate and silver chloride, and their effectiveness as biocides is widely known (Baker et al. 2005; Panáček et al. 2006; Makhluf et al. 2005). Though the antibacterial effect is size dependent, smaller Ag nanoparticles (8 nm) were most effective, while larger particle size (11–23 nm) resulted in lower bactericidal activity (Xiu et al. 2011). There are several mechanisms which have been found to explain the bactericidal effects of Ag nanoparticles such as, the damaging of the bacterial membranes due to the formation of free radicals (Xiu et al. 2012; Esteban-Cubillo et al. 2006) interactions with DNA, alteration of membrane properties due to adhesion on cell surface and enzyme damage.

Immobilized nanoparticles possess high antimicrobial activity (Balogh et al. 2001). In a study, cellulose acetate fibres embedded with Ag nanoparticles by direct electro spinning method were shown effective against both Gram-positive and Gram-negative bacteria. Ag nanoparticles are also incorporated into different types of polymers for the production of antimicrobial nanofibres (Chen et al. 2003; Botes and Cloete 2010; Chou et al. 2005). Water filters prepared using polyurethane foam coated with Ag nanofibres have shown good antibacterial properties against *Escherichia coli* (*E. coli*) (Lee et al. 2007).

Ag nanoparticles also find their applications in construction of filtration membranes, for example, in polysulphone membranes, for biofouling reduction. These

Ag nanoparticles laden membranes had good antimicrobial activities against *E. coli*, *Pseudomonas* and so forth (Adesina 2004; Li et al. 2008). Although Ag nanoparticles have been used efficiently for inactivating bacteria and viruses as well as reducing membrane biofouling, in the long term their capacity against membrane biofouling decreased considerably due to the loss of silver ions with time. Therefore, for long-term control of membrane biofouling, further work to reduce this loss of silver ions is required. Alternatively, doping of Ag nanoparticles with other metallic nanoparticles or its composites with metal oxide nanoparticles can solve the issue and this could also lead to the parallel removal of inorganic/organic compounds from wastewater.

8.1.2 TiO_2 Nanoparticles

TiO_2 nanoparticles are among the emerging and most promising photocatalysts for water purification. TiO_2 exposed to 8 h of simulated solar light, has been reported to reduce the viability of aqueous microbial pathogens, mainly protozoa and fungi. Nitrogen-doped TiO_2 nanoparticles catalysts have proved their efficiency for reduction in microbial contaminants in wastewater (Chaturvedi et al. 2012). Nanostructured TiO_2 films and membranes are capable of both disinfecting micro-organisms and decomposing organic pollutants under UV and visible light irradiation (Choi et al. 2009). Due to its stability in water, TiO_2 can be incorporated in thin films or merged onto membrane filters.

8.2 *Removal of Heavy Metals Ions*

The major heavy metals and metal ions present in wastewater from textile industries include copper, arsenic, lead, cadmium, mercury and chromium. Different types of nanomaterials have been introduced for removal of heavy metal ions from wastewater such as nanosorbents including CNTs, zeolites and dendrimers, and they have exceptional adsorption properties (Savage and Diallo 2005) The ability of CNTs to adsorb heavy metals such as Cd^{2+} (Li et al. 2003), Cr^{3+} (Di et al. 2006), Pb^{2+} (Li et al. 2005), and Zn^{2+} (Lu et al. 2006) and metalloids such as arsenic (As) compounds (Peng et al. 2005) has been well documented. CNT composites with iron and cerium oxide (CeO_2) have been used for the removal of heavy metal ions in few studies (Lu et al. 2005). CeO_2 nanoparticles supported with CNTs are used effectively to adsorb arsenic. Fast adsorption kinetics of CNTs is predominantly to the easily accessible adsorption sites and small intraparticle diffusion distance.

Metal-based nanomaterials are better in removing heavy metals than activated carbon (Sharma et al. 2009), for example, adsorption of arsenic by using TiO_2 nanoparticles and nanosized magnetite. Also, a nanocomposite of TiO_2 nanoparticles anchored on graphene sheet was used to reduce Cr(VI) to Cr

(III) in sunlight (Zhang et al. 2012). Similar Cr treatment was carried out by using palladium nanoparticles in another experiment (Omole et al. 2009). The capability of removing heavy metals such as As has also been investigated by using iron oxide nanomaterials (Fe_2O_3 and Fe_3O_4) as cost-effective adsorbents.

Iron nanoparticles (Fe^0) are very effective in reducing heavy metals such as As (III), As(V), Pb(II), Cu(II), Ni(II) and Cr(VI) (Ponder et al. 2000; Kanel et al. 2005, 2006; Yang and Lee 2005; Li and Zhang 2006). Novel self-assembled iron oxide nanostructures were also used to successfully adsorb both As(V) and Cr(VI) (Zhong et al. 2006). The 3D nanostructures of CeO_2 are used as good adsorbents for both As and Cr. Finally, there are commercial products for efficient removal of arsenic and these include FeO_2 nanoparticles and polymers and nanocrystalline TiO_2 medium as beads (Sylvester et al. 2007).

8.3 Removal of Organic Contaminants

Natural organic matter (NOM) constitutes a diverse group of hydrophobic (humic and fulvic acids) and hydrophilic organic compounds and it contributes significantly towards water contamination. Adsorption is the major principle used for removal of NOM from wastewater effluent and carbon-based adsorbents have been widely used for the same.

8.3.1 CNTs

Nanosorption has been carried out using nanosorbents, namely zeolites, polymeric materials such as dendrimers and carbon nanotubes (CNTs). These nanosorbents have exceptional adsorption properties and are used for removing organic content from wastewater (Savage and Diallo 2005). CNTs have exceptional water treatment capabilities due to their ability to adsorb high amount of organics on their surface (Rao et al. 2007) The removal of NOM by CNTs is higher in comparison with other carbon-based adsorbents because of the provision of large surface area in CNTs (Saleh et al. 2008). CNTs are also effective in removing polycyclic aromatic compounds (Di et al. 2006) and hence find special application in the textile industry as the dyeing process releases maximum amount of aromatic compounds into the effluent stream. Activated carbon fibres prepared by electrospinning of CNTs which are nanoporous in nature showed much greater organic sorption for benzene, toluene, xylene and ethyl benzene (which are used as solvents in the textile industry) than granular activated carbon (Mangun et al. 2001). Multi-walled CNTs functionalized with Fe nanoparticles can be used as effective sorbents for aromatic compounds such as benzene and toluene (Jin et al. 2007).

8.3.2 TiO_2 Nanoparticles

Nanomaterials of metal oxides such as TiO_2 in addition to CeO_2 have also been used as catalysts for quick and relatively higher degree of degradation of organic contaminants in ozonation processes (Nawrocki and Kasprzyk-Hordern 2010; Orge et al. 2011). Polychlorinated biphenyls (PCBs) which are highly toxic to the environment are majorly used in nylon industries and can enter the effluent stream during discharge. Photocatalysts such as TiO_2 nanoparticles are used effectively for treating wastewater polluted by organic contaminants such as polychlorinated biphenyls (PCBs), benzenes and chlorinated alkanes (Kabra et al. 2004). Removal of total organic carbon from wastewater has been enhanced by the use of TiO_2.

Decomposition of organic compounds can be enhanced by noble metal doping into TiO_2 due to enhanced hydroxyl radical production and so forth. For example, the doping of Si into TiO_2 nanoparticles was proved to be effectual in improving its efficiency due to the increase in surface area and crystallinity (Iwamoto et al. 2000, 2005). TiO_2 nanocrystals modified with noble metal deposits and so forth were used for degrading methylene blue dye in the visible light range (Liu and Gao 2005; Wu et al. 2006). Nitrogen- and Fe(III)-doped TiO_2 nanoparticles are useful in degrading azo dyes and phenol, respectively, than commercially available TiO_2 nanoparticles. TiO_2 nanotubes have been effectively used to degrade organic compounds and were found to be more efficient than TiO_2 nanoparticles (Macak et al. 2007).

8.3.3 Zero-Valent Iron

Azo dyes are among the pollutants of significant concern that are being let out by most of the textile industries. Nanocatalysts including semiconductors, zero-valent metals, and bimetallic nanoparticles have been used for the removal of several organic pollutants, namely PCBs, pesticides and azo dyes. The main property that enables them to be effective is their greater surface area and shape specific properties (Zhao et al. 2011).

Other class of chlorinated organic compounds and PCBs have been transformed successfully using nanoscale zero-valent iron (nZVI) (Wang and Zhang 1997) and inorganic ions such as nitrate and perchlorate (Choe et al. 2000; Cao et al. 2005). The stabilized nZVI particles could also be an efficacious way for in situ treatment of industrial effluents such as the textile industry (Xu and Zhao 2007). The nZVI and bimetallic nZVI can be used as effective redox reagents for reducing a variety of organic pollutants released by textile industries such as PCBs and organic dyes owing to larger surface areas and higher reactivity (Schrick et al. 2002; Nurmi et al. 2005).

8.4 Other Nanomaterials

The nanocatalyst of Ag and amidoxime fibres was used efficiently for the degradation of organic dyes useful for treating effluent from textile industries (Wu et al. 2010). Manganese dioxide (MnO_2) films have been used for the mineralization of organic dyes (Espinal et al. 2004). Similarly, MnO_2 hierarchical hollow nanostructures have been put to use for the removal of organic pollutant in wastewater (Fei et al. 2008). Several millions of dollars are invested into the material research of such useful nanochemicals.

9 Challenges and Limitations of Using Nanochemicals

The uses of nanochemicals for wastewater treatment remains to be relatively less explored. Several academic breakthroughs and research are yet to be translated to benefit the industry and their allied applications. The use of nanochemicals for photocatalysis and nanosorption for the wastewater treatment in textile industries requires skilled personnel with particular expertise in the area. Multiple modifications in the existing treatment equipment are also necessary for the application of nanochemicals to be functional and efficient. The lack of knowledge on the fate of nanochemicals in water and the difficulty to predict their properties and behaviour is of concern. Unintended pollution–catalytic effects on the environment through reactions in macro-interfacial processes accounts as a potential hazard (Yao et al. 2013).

10 Conclusion

The textile industry consists of a series of processes which lead to discharge of harmful pollutants into the effluent stream. These pollutants if released unchecked and untreated can cause adverse effects to the environment and aquatic life. Hence, the effluents from textile industries need to be properly treated and discharged. Present techniques are finding it difficult to keep up with effluent standards and volumes of pollutants being released. Novel techniques are being developed and one of the most promising solutions could be the use of nanochemicals for the treatment of wastewater from textile industries.

Nanochemicals are capable of removal of pollutants by photocatalysis or nanosorption. They can also be moulded to form nanofiltration membranes. Nanochemicals can be used for the removal of organic dyes released throughout the dyeing process, solvents released during the scouring process, heavy metals and ions released during both the dyeing and printing process. TiO_2, nZVI, zeolites are some of the nanochemicals which are of great use and can be used practically for the treatment of wastewater effluent from textile industries.

References

Abdolmohammad-Zadeh, H., Ghorbani, E., & Talleb, Z. (2013). Zinc–aluminum layered double hydroxide as a nano-sorbent for removal of Reactive Yellow 84 dye from textile wastewater effluents. *Journal of the Iranian Chemical Society, 10*, 1103–1112.

Abraham, T. E., Senan, R. C., Shaffiqu, T. S., Roy, J. J., Poulose, T. P., & Thomas, P. P. (2003). Bioremediation of textile azo dyes by an aerobic bacterial consortium using a rotating biological contactor. *Biotechnology Progress, 19*, 1372–1376.

Adesina, A. A. (2004). Industrial exploitation of photocatalysis: Progress, perspectives and prospects. *Catalysis Surveys from Asia, 10*, 265–273.

Alivisatos, A. P. (1996). Perspectives on the physical chemistry of semiconductor nanocrystals. *The Journal of Physical Chemistry, 100*, 13226–13239.

Allabashi, R., Arkas, M., Hörmann, G., & Tsiourvas, D. (2007). Removal of some organic pollutants in water employing ceramic membranes impregnated with cross-linked silylated dendritic and cyclodextrin polymers. *Water Research, 41*, 476–486.

Al-Omair, M. A., & El-Sharkawy, E. A. (2007). Removal of heavy metals via adsorption on activated carbon synthesized from solid wastes. *Environmental Technology, 28*, 443–451.

Asenjo, N. G., Álvarez, P., Granda, M., Blanco, C., Santamaría, R., & Menéndez, R. (2011). High performance activated carbon for benzene/toluene adsorption from industrial wastewater. *Journal of Hazardous Materials, 192*, 1525–1532.

Babu, B. R., Parande, A. K., & Raghu, S. (2007). Cotton textile processing: Waste generation and effluent treatment. *Journal of cotton science, 11*, 141–153.

Baker, C., Pradhan, A., Pakstis, L., Pochan, D. J., & Shah, S. I. (2005). Synthesis and antibacterial properties of silver nanoparticles. *Journal of Nanoscience and Nanotechnology, 5*, 244–249.

Balogh, L., Swanson, D. R., Tomalia, D. A., Hagnauer, G. L., & McManus, A. T. (2001). Dendrimer-silver complexes and nanocomposites as antimicrobial agents. *Nano Letters, 1*, 18–21.

Barhate, R. S., & Ramakrishna, S. (2007). Nanofibrous filtering media: Filtration problems and solutions from tiny materials. *Journal of Membrane Science, 296*, 1–8.

Baruah, S., & Dutta, J. (2016). Hydrothermal growth of ZnO nanostructures. *Science and Technology of Advanced Materials*.

Baruah, S., K Pal, S., & Dutta, J. (2012). Nanostructured zinc oxide for water treatment. *Nanoscience & Nanotechnology-Asia, 2*, 90–102.

Berry, D., Xi, C., & Raskin, L. (2006). Microbial ecology of drinking water distribution systems. *Current Opinion in Biotechnology, 17*, 297–302.

Bhatkhande, D. S., Pangarkar, V. G., & Beenackers, A. A. (2002). Photocatalytic degradation for environmental applications–a review. *Journal of Chemical Technology and Biotechnology, 77*, 102–116.

Borgarello, E., Serpone, N., Emo, G., Harris, R., Pelizzetti, E., & Minero, C. (1986). *Inorganic Chemistry, 25*, 4499.

Botes, M., & Cloete, T. E. (2010). The potential of nanofibers and nanobiocides in water purification. *Critical Reviews in Microbiology, 36*, 68–81.

Cao, J., Elliott, D., & Zhang, W. X. (2005). Perchlorate reduction by nanoscale iron particles. *Journal of Nanoparticle Research, 7*, 499–506.

Chae, S. R., Wang, S., Hendren, Z. D., Wiesner, M. R., Watanabe, Y., & Gunsch, C. K. (2009). Effects of fullerene nanoparticles on *Escherichia coli* K12 respiratory activity in aqueous suspension and potential use for membrane biofouling control. *Journal of Membrane Science, 329*, 68–74.

Chaohai, W., Xiangdong, J., & Huanqin, C. (1998). Advances in the technology of wastewater treatment by aerobic biological fluidized bed. *Journal of Environmental Science and Technology, 4*, 001.

Chaturvedi, S., Dave, P. N., & Shah, N. K. (2012). Applications of nano-catalyst in new era. *Journal of Saudi Chemical Society, 16*, 307–325.

Chen, Y., Wang, L., Jiang, S., & Yu, H. (2003). Study on novel antibacterial polymer materials (I) preparation of zeolite antibacterial agents and antibacterial polymer composite and their antibacterial properties. *Journal of Polymer Materials, 20*, 279–284.

Choe, S., Chang, Y. Y., Hwang, K. Y., & Khim, J. (2000). Kinetics of reductive denitrification by nanoscale zero-valent iron. *Chemosphere, 41*, 1307–1311.

Choi, H., Al-Abed, S. R., & Dionysiou, D. D. (2009). Nanostructured titanium oxide film- and membrane-based photocatalysis for water treatment. *Nanotechnology Applications for Clean Water, 3*, 39–46.

Chou, G. E. W. L., Yu, D. G., & Yang, M. C. (2005). The preparation and characterization of silver-loading cellulose acetate hollow fibre membrane for water treatment. *Polymers for Advanced Technologies, 16*, 600–607.

Crane, R. A., & Scott, T. B. (2012). Nanoscale zero-valent iron: Future prospects for an emerging water treatment technology. *Journal of Hazardous Materials, 211*, 112–125.

Dabrowski, A., Hubicki, Z., Podkościelny, P., & Robens, E. (2004). Selective removal of the heavy metal ions from waters and industrial wastewaters by ion-exchange method. *Chemosphere, 56*, 91–106.

Dalan, J. A. (2000). 9 things to know about zero liquid discharge. *Chemical Engineering Progress, 96*, 71–76.

Delée, W., O'Neill, C., Hawkes, F. R., & Pinheiro, H. M. (1998). Anaerobic treatment of textile effluents: A review. *Journal of Chemical Technology and Biotechnology, 73*, 323–335.

Di, Z. C., Ding, J., Peng, X. J., Li, Y. H., Luan, Z. K., & Liang, J. (2006). Chromium adsorption by aligned carbon nanotubes supported ceria nanoparticles. *Chemosphere, 62*, 861–865.

Dugan, N. R., & Williams, D. J. (2006). Cyanobacteria passage through drinking water filters during perturbation episodes as a function of cell morphology, coagulant and initial filter loading rate. *Harmful Algae, 5*, 26–35.

Eggins, B. R., Palmer, F. L., & Byrne, J. A. (1997). Photocatalytic treatment of humic substances in drinking water. *Water Research, 31*, 1223–1226.

Eijkel, J. C. T., & Van den Berg, A. (2005). Nanofluidics: What is it and what can we expect from it? *Microfluidics and Nanofluidics, 1*, 249–267.

El-Safty, S. A., Shahat, A., & Ismael, M. (2012). Mesoporous aluminosilica monoliths for the adsorptive removal of small organic pollutants. *Journal of Hazardous Materials, 201*, 23–32.

Escobar, I. C., Randall, A. A., & Taylor, J. S. (2001). Bacterial growth in distribution systems: Effect of assimilable organic carbon and biodegradable dissolved organic carbon. *Environmental Science and Technology, 35*, 3442–3447.

Espinal, L., Suib, S. L., & Rusling, J. F. (2004). Electrochemical catalysis of styrene epoxidation with films of MnO_2 nanoparticles and H_2O_2. *Journal of the American Chemical Society, 126*, 7676–7682.

Esteban-Cubillo, A., Pecharromán, C., Aguilar, E., Santarén, J., & Moya, J. S. (2006). Antibacterial activity of copper monodispersed nanoparticles into sepiolite. *Journal of Materials Science, 41*, 5208–5212.

Fei, J., Cui, Y., Yan, X., et al. (2008). Controlled preparation of MnO_2 hierarchical hollow nanostructures and their application in water treatment. *Advanced Materials, 20*, 452–456.

Fritzmann, C., Löwenberg, J., Wintgens, T., & Melin, T. (2007). State-of-the-art of reverse osmosis desalination. *Desalination, 216*, 1–76.

Guosheng, C. H. S. (2000). Research progress and application on biological contact oxidation of micropolluted source water. *Techniques and Equipment for Environmental Pollution, 3*, 10.

Heijman, S. G. J., Guo, H., Li, S., Van Dijk, J. C., & Wessels, L. P. (2009). Zero liquid discharge: Heading for 99 % recovery in nanofiltration and reverse osmosis. *Desalination, 236*, 357–362.

Hoffmann, M. R., Martin, S. T., Choi, W., & Bahnemann, D. W. (1995). *Chemical Reviews, 95*, 69.

Hornyak, G. L., Dutta, J., Tibbals, H. F., & Rao, A. (2008). *Introduction to nanoscience*. Boca Raton: CRC Press.

Hornyak, G. L., Moore, J. J., Tibbals, H. F., & Dutta, J. (2009). *Fundamentals of nanotechnology*. Boca Raton: CRC Press.

Hu, H., Wang, Z., & Pan, L. (2010). Synthesis of monodisperse Fe_3O_4@ silica core–shell microspheres and their application for removal of heavy metal ions from water. *Journal of Alloys and Compounds, 492*, 656–661.

Ibrahim, K., & Khoury, H. (2002). Use of natural chabazite–phillipsite tuff in wastewater treatment from electroplating factories in Jordan. *Environmental Geology, 41*, 547–551.

Iram, M., Guo, C., Guan, Y., Ishfaq, A., & Liu, H. (2010). Adsorption and magnetic removal of neutral red dye from aqueous solution using Fe_3O_4 hollow nanospheres. *Journal of Hazardous Materials, 181*, 1039–1050.

Iwamoto, S., Iwamoto, S., Inoue, M., Yoshida, H., Tanaka, T., & Kagawa, K. (2005). XANES and XPS study of silica-modified titanias prepared by the glycothermal method. *Chemistry of Materials, 17*, 650–655.

Iwamoto, S., Tanakulrungsank, W., Inoue, M., Kagawa, K., & Praserthdam, P. (2000). Synthesis of large-surface area silica-modified Titania ultrafine particles by the glycothermal method. *Journal of Materials Science Letters, 19*, 1439–1443.

Jain, P., & Pradeep, T. (2005). Potential of silver nanoparticle-coated polyurethane foam as an antibacterial water filter. *Biotechnology and Bioengineering, 90*, 59–63.

Jeon, S., & Yong, K. (2010). Morphology-controlled synthesis of highly adsorptive tungsten oxide nanostructures and their application to water treatment. *Journal of Materials Chemistry, 20*, 10146–10151.

Jin, J., Li, R., Wang, H., Chen, H., Liang, K., & Ma, J. (2007). Magnetic Fe nanoparticle functionalized water-soluble multi-walled carbon nanotubules towards the preparation of sorbent for aromatic compounds removal. *Chemical Communications, 4*, 386–388.

Jiraratananon, R., Sungpet, A., & Luangsowan, P. (2000). Performance evaluation of nanofiltration membranes for treatment of effluents containing reactive dye and salt. *Desalination, 130*, 177–183.

Kabra, K., Chaudhary, R., & Sawhney, R. L. (2004). Treatment of hazardous organic and inorganic compounds through aqueous-phase photocatalysis: A review. *Industrial and Engineering Chemistry Research, 43*, 7683–7696.

Kadirvelu, K., Thamaraiselvi, K., & Namasivayam, C. (2001). Removal of heavy metals from industrial wastewaters by adsorption onto activated carbon prepared from an agricultural solid waste. *Bioresource Technology, 76*, 63–65.

Kanel, S. R., Greneche, J. M., & Choi, H. (2006). Arsenic(V) removal from groundwater using nano scale zero-valent iron as a colloidal reactive barrier material. *Environmental Science and Technology, 40*, 2045–2050.

Kanel, S. R., Manning, B., Charlet, L., & Choi, H. (2005). Removal of arsenic(III) from groundwater by nanoscale zero-valent iron. *Environmental Science and Technology, 39*, 1291–1298.

Kesraoui-Ouki, S., Cheeseman, C., & Perry, R. (1993). Effects of conditioning and treatment of chabazite and clinoptilolite prior to lead and cadmium removal. *Environmental Science and Technology, 27*, 1108–1116.

Kim, J., Davies, S. H. R., Baumann, M. J., Tarabara, V. V., & Masten, S. J. (2008). Effect of ozone dosage and hydrodynamic conditions on the permeate flux in a hybrid ozonation-ceramic ultrafiltration system treating natural waters. *Journal of Membrane Science, 311*, 165–172.

Kobya, M., Demirbas, E., Senturk, E., & Ince, M. (2005). Adsorption of heavy metal ions from aqueous solutions by activated carbon prepared from apricot stone. *Bioresource Technology, 96*, 1518–1521.

Kumar, P. S., Vijayakumar Abhinaya, R., Arthi, V., GayathriLashmi, K., Priyadharshini, M. & Sivanesan, S. (2014). Adsorption of methylene blue dye onto surface modified cashew nut shell. *Environmental Engineering & Management Journal (EEMJ), 13*.

Langlais, B., Reckhow, D. A., & Brink, D. R. (1991). *Ozone in water treatment: Application and engineering*. Boca Raton: CRC Press.

Lee, S. Y., Kim, H. J., Patel, R., Im, S. J., Kim, J. H., & Min, B. R. (2007). Silver nanoparticles immobilized on thin film composite polyamide membrane: Characterization, nanofiltration, antifouling properties. *Polymers for Advanced Technologies, 18*, 562–568.

Lee, J., Park, H., & Choi, W. (2002). Selective photocatalytic oxidation of NH_3 to N_2 on platinized TiO_2 in water. *Environmental Science and Technology, 36*, 5462–5468.

Leng, C. C., & Pinto, N. G. (1996). An investigation of the mechanisms of chemical regeneration of activated carbon. *Industrial and Engineering Chemistry Research, 35*, 2024–2031.

Li, Y. H., Di, Z., Ding, J., Wu, D., Luan, Z., & Zhu, Y. (2005). Adsorption thermodynamic, kinetic and desorption studies of $Pb2^{+}$ on carbon nanotubes. *Water Research, 39*, 605–609.

Li, Y. H., Ding, J., Luan, Z., et al. (2003). Competitive adsorption of Pb^{2+}, Cu^{2+} and Cd^{2+} ions from aqueous solutions by multiwalled carbon nanotubes. *Carbon, 41*, 2787–2792.

Li, Q., Mahendra, S., Lyon, D. Y., et al. (2008). Antimicrobial nanomaterials for water disinfection and microbial control: Potential applications and implications. *Water Research, 42*, 4591–4602.

Li, J. F., Xu, Z. L., Yang, H., Yu, L. Y., & Liu, M. (2009). Effect of TiO_2 nanoparticles on the surface morphology and performance of microporous PES membrane. *Applied Surface Science, 255*, 4725–4732.

Li, X. Q., & Zhang, W. X. (2006). Iron nanoparticles: The core-shell structure and unique properties for Ni(II) sequestration. *Langmuir, 22*, 4638–4642.

Liu, H., & Gao, L. (2005). Synthesis and properties of CdSe-sensitized rutile TiO_2 nanocrystals as a visible light-responsive photocatalyst. *Journal of the American Ceramic Society, 88*, 1020–1022.

Lu, C., Chiu, H., & Liu, C. (2006). Removal of zinc(II) from aqueous solution by purified carbon nanotubes: Kinetics and equilibrium studies. *Industrial and Engineering Chemistry Research, 45*, 2850–2855.

Lu, C., Chung, Y. L., & Chang, K. F. (2005). Adsorption of trihalomethanes from water with carbon nanotubes. *Water Research, 39*, 1183–1189.

Luo, L. H., Feng, Q. M., Wang, W. Q., & Zhang, B. L. (2011). Fe_3O_4/Rectorite composite: Preparation, characterization and absorption properties from contaminant contained in aqueous solution. *Advanced Materials Research, 287*, 592–598.

Macak, J. M., Zlamal, M., Krysa, J., & Schmuki, P. (2007). Self-organized TiO_2 nanotube layers as highly efficient photocatalysts. *Small (Weinheim an der Bergstrasse, Germany), 3*, 300–304.

Majsterek, I., Sicinska, P., Tarczynska, M., Zalewski, M., & Walter, M. (2004). Toxicity of microcystin from cyanobacteria growing in a source of drinking water. *Comparative Biochemistry and Physiology—C Toxicology and Pharmacology, 139*, 175–179.

Makhluf, S., Dror, R., Nitzan, Y., Abramovich, Y., Jelinek, R., & Gedanken, A. (2005). Microwave-assisted synthesis of nanocrystalline MgO and its use as a bacteriocide. *Advanced Functional Materials, 15*, 1708–1715.

Maliou, E., Malamis, M., & Sakellarides, P. O. (1992). Lead and cadmium removal by ion exchange. *Water Science and Technology, 25*, 133–138.

Mangun, C. L., Yue, Z., Economy, J., Maloney, S., Kemme, P., & Cropek, D. (2001). Adsorption of organic contaminants from water using tailored ACFs. *Chemistry of Materials, 13*, 2356–2360.

Masciangioli, J., & Zhang, W. X. (2003). Peer reviewed: Environmental technologies at the nanoscale. *Environmental Science and Technology, 37*, 102–108.

McNaught, A. D., & Wilkinson, A. (1997). *IUPAC. Compendium of chemical terminology ("gold book")* (2nd ed.). Oxford: Blackwell Scientific Publications. (XML on-line corrected version created by Nic M, Jirat J, Kosata B).

Menezes, E., & Choudhari, M. (2011). *Pre-treatment of textiles prior to dyeing.*

Mills, A., Belghazi, A., & Rodman, D. (1996). Bromate removal from drinking water by semiconductor photocatalysis. *Water Research, 30*, 1973–1978.

Mills, A., Davies, R. H., & Worsley, D. (1993). Water purification by semiconductor photocatalysis. *Chemical Society Reviews, 22*, 417–425.

Nassar, N. N. (2010). Rapid removal and recovery of Pb (II) from wastewater by magnetic nanoadsorbents. *Journal of Hazardous Materials, 184*, 538–546.

Nawrocki, J., & Kasprzyk-Hordern, B. (2010). The efficiency and mechanisms of catalytic Ozonation. *Applied Catalysis, B: Environmental, 99*, 27–42.

Nguyen, T., Roddick, F. A., & Fan, L. (2012). Biofouling of water treatment membranes: A review of the underlying causes, monitoring techniques and control measures. *Membranes, 2*, 804–840.

Nurmi, J. T., Tratnyek, P. G., Sarathy, V., et al. (2005). Characterization and properties of metallic iron nanoparticles: Spectroscopy, electrochemistry, and kinetics. *Environmental Science and Technology, 39*, 1221–1230.

Oliveira, L. C. A, Rachel Rios, V. R. A., Jose Fabris, D., Garg, V., Karim Sapag, & Rochel Lago, M. (2002). Activated carbon/iron oxide magnetic composites for the adsorption of contaminants in water. *Carbon, 40*, 2177–2183.

Omole, M. A., K'Owino, I., & Sadik, O. A. (2009). Nanostructured materials for improving water quality: Potentials and risks. *Nanotechnology Applications for Clean Water, 17*, 233–247.

Orge, C. A., Órfão, J. J. M., Pereira, M. F. R., Duarte de Farias, A. M., Neto, R. C. R., & Fraga, M. A. (2011). Ozonation of model organic compounds catalysed by nanostructured cerium oxides. *Applied Catalysis, B: Environmental, 103*, 190–199.

Ouki, S. K., & Kavannagh, M. (1997). Performance of natural zeolites for the treatment of mixed metal-contaminated effluents. *Waste Management and Research, 15*, 383–394.

Pala, A., & Tokat, E. (2002). Color removal from cotton textile industry wastewater in an activated sludge system with various additives. *Water Research, 36*, 2920–2925.

Panáček, A., Kvítek, L., Prucek, R., et al. (2006). Silver colloid nanoparticles: Synthesis, characterization, and their antibacterial activity. *The Journal of Physical Chemistry B, 110*, 16248–16253.

Pansini, M., Colella, C., & De Gennaro, M. (1991). Chromium removal from water by ion exchange using zeolite. *Desalination, 83*, 145–157.

Peng, X., Luan, Z., Ding, J., Di, Z., Li, Y., & Tian, B. (2005). Ceria nanoparticles supported on carbon nanotubes for the removal of arsenate from water. *Materials Letters, 59*, 399–403.

Pirkanniemi, K., & Sillanpää, M. (2002). Heterogeneous water phase catalysis as an environmental application: A review. *Chemosphere, 48*, 1047–1060.

Ponder, S. M., Darab, J. G., & Mallouk, T. E. (2000). Remediation of Cr(VI) and Pb(II) aqueous solutions using supported, nanoscale zero-valent iron. *Environmental Science and Technology, 34*, 2564–2569.

Rai, H. S., Bhattacharyya, M. S., Singh, J., Bansal, T. K., Vats, P., & Banerjee, U. C. (2005). Removal of dyes from the effluent of textile and dyestuff manufacturing industry: A review of emerging techniques with reference to biological treatment. *Critical reviews in environmental science and technology, 35*, 219–238.

Rao, G. P., Lu, C., & Su, F. (2007). Sorption of divalent metal ions from aqueous solution by carbon nanotubes: A review. *Separation and Purification Technology, 58*, 224–231.

Razzak, N. R. B. (2014). *Effectiveness of fenton's reagent in the treatment of textile effluent.*

Rickerby, D. G., & Morrison, M. (2007). Nanotechnology and the environment: A European perspective. *Science and Technology of Advanced Materials, 8*, 19–24.

Riu, J., Maroto, A., & Rius, F. X. (2006). Nanosensors in environmental analysis. *Talanta, 69*, 288–301.

Saleh, N. B., Pfefferle, L. D., & Elimelech, M. (2008). Aggregation kinetics of multiwalled carbon nanotubes in aquatic systems: Measurements and environmental implications. *Environmental Science and Technology, 42*, 7963–7969.

Savage, N., & Diallo, M. S. (2005). Nanomaterials and water purification: Opportunities and challenges. *Journal of Nanoparticle Research, 7*, 331–342.

Saxena, S., & Kaushik, S. (2011). *Effluent treatment in textile industries.*

Schrick, B., Blough, J. L., Jones, A. D., & Mallouk, T. E. (2002). Hydrodechlorination of trichloroethylene to hydrocarbons using bimetallic nickel-iron nanoparticles. *Chemistry of Materials, 14*, 5140–5147.

Serpone, N., Ah-You, Y. K., Tran, T. P., Harris, R., Pelizzetti, E., & Hidaka, H. (1987). AM1 simulated sunlight photoreduction and elimination of Hg (II) and CH_3 Hg (II) chloride salts from aqueous suspensions of titanium dioxide. *Solar Energy, 39*, 491–498.

Sharma, Y. C., Srivastava, V., Singh, V. K., Kaul, S. N., & Weng, C. H. (2009). Nano-adsorbents for the removal of metallic pollutants from water and wastewater. *Environmental Technology, 30*, 583–609.
Sondi, I., & Salopek-Sondi, B. (2004). Silver nanoparticles as antimicrobial agent: A case study on *E. coli* as a model for Gram-negative bacteria. *Journal of Colloid and Interface Science, 275*, 177–182.
Srinivasan, S., Harrington, G. W., Xagoraraki, I., & Goel, R. (2008). Factors affecting bulk to total bacteria ratio in drinking water distribution systems. *Water Research, 42*, 3393–3404.
Sylvester, P., Westerhoff, P., Möller, T., Badruzzaman, M., & Boyd, O. (2007). A hybrid sorbent utilizing nanoparticles of hydrous iron oxide for arsenic removal from drinking water. *Environmental Engineering Science, 24*, 104–112.
Theron, J., Walker, W. A., & Cloete, T. E. (2008). Nanotechnology and water treatment: Applications and emerging opportunities. *Critical Reviews in Microbiology, 34*, 43–69.
Tüfekci, Neşe, Sivri, Nüket, & Toroz, İsmail. (2007). Pollutants of textile industry wastewater and assessment of its discharge limits by water quality standards. *Turkish Journal of Fisheries and Aquatic Sciences, 7*, 97–103.
Vaseashta, A., Vaclavikova, M., Vaseashta, S., Gallios, G., Roy, P., & Pummakarnchana, O. (2007). Nanostructures in environmental pollution detection, monitoring, and remediation. *Science and Technology of Advanced Materials, 8*, 47–59.
Verma, A. K., Dash, R. R., & Bhunia, P. (2012). A review on chemical coagulation/flocculation technologies for removal of colour from textile wastewaters. *Journal of Environmental Management, 93*, 154–168.
Wang, Z., Huang, K., Xue, M., & Liu, Z. (2011). *Textile dyeing wastewater treatment,* 91–116.
Wang, C. B., & Zhang, W. X. (1997). Synthesizing nanoscale iron particles for rapid and complete dechlorination of TCE and PCBs. *Environmental Science and Technology, 31*, 2154–2156.
Wayman, J. (2015). Brackish ground water desalination using solar reverse osmosis.
Wu, L., Yu, J. C., & Fu, X. (2006). Characterization and photocatalytic mechanism of nanosized CdS coupled TiO_2 nanocrystals under visible light irradiation. *Journal of Molecular Catalysis A: Chemical, 244*, 25–32.
Wu, Z. C., Zhang, Y., Tao, T. X., Zhang, L., & Fong, H. (2010). Silver nanoparticles on amidoxime fibres for photo-catalytic degradation of organic dyes in waste water. *Applied Surface Science, 257*, 1092–1097.
Xiu, Z. M., Ma, J., & Alvarez, P. J. J. (2011). Differential effect of common ligands and molecular oxygen on antimicrobial activity of silver nanoparticles versus silver ions. *Environmental Science and Technology, 45*, 9003–9008.
Xiu, Z. M., Zhang, Q. B., Puppala, H. L., Colvin, V. L., & Alvarez, P. J. J. (2012). Negligible particle-specific antibacterial activity of silver nanoparticles. *Nano Letters, 12*, 4271–4275.
Xu, Y., & Zhao, D. (2007). Reductive immobilization of chromate in water and soil using stabilized iron nanoparticles. *Water Research, 41*, 2101–2108.
Yang, G. C. C., & Lee, H. L. (2005). Chemical reduction of nitrate by nanosized iron: Kinetics and pathways. *Water Research, 39*, 884–894.
Yantasee, W., Rutledge, R. D., Chouyyok, W., Sukwarotwat, V., Orr, G., Warner, C. L., et al. (2010). Functionalized nanoporous silica for the removal of heavy metals from biological systems: Adsorption and application. *ACS Applied Materials & Interfaces, 2*, 2749–2758.
Yao, D., Chen, Z., Zhao, K., Yang, Q., & Zhang, W. (2013). Limitation and challenge faced to the researches on environmental risk of nanotechnology. *Procedia Environmental Sciences, 18*, 149–156.
Zamzow, M. J., Eichbaum, B. R., Sandgren, K. R., & Shanks, D. E. (1990). Removal of heavy metals and other cations from wastewater using zeolites. *Separation Science and Technology, 25*, 1555–1569.
Zhang, K., Kemp, K. C., & Chandra, V. (2012). Homogeneous anchoring of TiO_2 nanoparticles on graphene sheets for waste water treatment. *Materials Letters, 81*, 127–130.
Zhang, S., Niu, H., Hu, Z., Cai, Y., & Shi, Y. (2010). Preparation of carbon coated Fe_3O_4 nanoparticles and their application for solid-phase extraction of polycyclic aromatic

hydrocarbons from environmental water samples. *Journal of Chromatography A, 1217*, 4757–4764.

Zhang, S., Xu, W., Zeng, M., Li, J., Li, J., Xu, J., et al. (2013). Superior adsorption capacity of hierarchical iron oxide @ magnesium silicate magnetic nanorods for fast removal of organic pollutants from aqueous solution. *Journal of Materials Chemistry A, 1*, 11691–11697.

Zhao, X., Lv, L., Pan, B., Zhang, W., Zhang, S., & Zhang, Q. (2011). Polymer-supported nanocomposites for environmental application: A review. *Chemical Engineering Journal, 170*, 381–394.

Zhong, L. S., Hu, J. S., Liang, H. M., Cao, A. M., Song, W. G., & Wan, L. J. (2006). Self-assembled 3D flowerlike iron oxide nanostructures and their application in water treatment. *Advanced Materials, 18*, 2426–2431.

Insights into the Functional Finishing of Textile Materials Using Nanotechnology

Shahid-ul-Islam, Mohd Shabbir and Faqeer Mohammad

Abstract Over the past few decades, there is an emergence of new multidisciplinary approaches to functionalize different textile materials. Nanotechnology is increasingly attracting scientific attention to develop multifunctional textiles for various end uses among all technologies. Nanoparticles play vital role in coloration and, in view of their large surface area-to-volume ratio and high surface energy, have imparted novel properties such as microbial resistance, flame retardancy, and self-cleaning property to different textile surfaces. This book chapter emphasizes on recent functional treatments of both natural and synthetic textile materials using nanotechnology. Applications of the sustainable nanotextiles in many of the sectors such as medicine and protective clothing are also critically discussed.

Keywords Nanotextiles · Sustainable clothing · Functionalization · Antimicrobial activity

1 Introduction

Textile and clothing industries play major role in the economy of many countries. Clothing has been a basic need besides food and home from the beginning of this world. Regular modernization of the society developed the clothing sense of the human being for different time periods (Islam et al. 2014). Earlier, clothes were used to cover the body or to look good, and now because of the increased consciousness about better life, consumers prefer garments that show multifunctional properties (Montazer et al. 2013; Shahid et al. 2013). Clothing today such as wool, cotton, silk, and other synthetic fibers is

---

Shahid-ul-Islam (✉) · M. Shabbir · F. Mohammad
Department of Chemistry, Jamia Millia Islamia, New Delhi 110025, India
e-mail: shads.jmi@gmail.com

M. Shabbir
e-mail: shabbirmeo@gmail.com

F. Mohammad
e-mail: faqeermohammad@rediffmail.com

S.S. Muthu (ed.), *Textiles and Clothing Sustainability*,
Textile Science and Clothing Technology,
DOI 10.1007/978-981-10-2188-6_3

expected to be stain, water, and insect repellent, antibacterial, and UV protective. (Kaplan and Okur 2008). Textile fibers functionalized with different agents are now widely used in different application sectors such as textile, pharmaceutical, medical, engineering, agricultural, and food industries. As a consequence of their importance, many synthetic antimicrobial agents such as triclosan, quaternary ammonium compounds, *N*-halamines, dyes, metal salts, and other natural product-derived agents including tea polyphenols, natural dyes, and chitosan have been extensively used (Hutchison 2008; Gao and Cranston 2008; Simoncic and Tomsic 2010). Textile coloration and functional finishing is a wet process consuming lot of chemicals, water, and synthetic dyes (Batool et al. 2013; Islam et al. 2013a, b). Scientists are working to discover new innovative technologies, and old processes and techniques are constantly being replaced by new ones. In view of the rising environmental pollution concerns, not only the methods but environmentally friendly products are also introduced in dyeing and finishing industries (Shahid et al. 2013; Islam et al. 2014). Among various sustainable and renewable products, the application of natural dyes extracted from plant species such as annatto, gallnut, *Rheum emodi*, henna, walnut, madder, pomegranate, cutch, babool, onion, and carrot has been the subject of several recent investigations aimed at obtaining color as well as other functional characteristics (Bhatti et al. 2010; Khan et al. 2012).

Lately, researchers have shown much more interest in using nanotechnology for functional modification of different natural and synthetic textile materials. Research is underway worldwide on different nanomaterials to impart some new prominent properties to fabrics. This chapter first discusses some conventional methods and

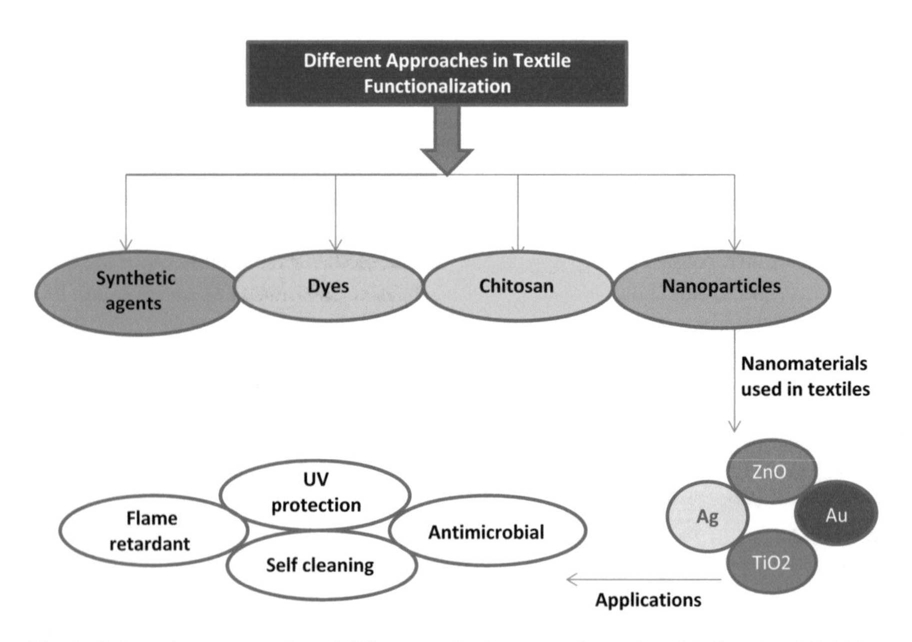

Fig. 1 Schematic representation of different methods commonly employed in functional finishing of textile materials

finally highlights the role of nanotechnology in functional modification of different textile materials (Fig. 1).

2 Conventional Methods for Functional Finishing

A number of approaches and strategies are at operation to produce textile surfaces having antimicrobial, UV protection, antioxidant, and other hygienic properties. The most important conventional methods used for functionalization of textiles are described below.

2.1 Synthetic Agents for Functional Modification

Extensive investigations are reported over the past few decades to study the application of different chemicals to textile surfaces for antibacterial, UV protection, and flame-retardant finishing (Montazer et al. 2013). Among various functional finishing methods, antimicrobial treatments are given much more importance as they pertain to precious human life. Textile materials are well suited for bacterial growth and therefore act as potential vectors to transmit diseases. It is recommended that antimicrobial modification of textiles should be durable to washing; dry-cleaning and hot pressing and at the same time the quality or appearance of the

Table 1 Some commercially available antimicrobial agents (Gao and Cranston 2008; Simoncic and Tomsic 2010)

Antimicrobial agent	Chemical formula	Structure
Chlorohexidine	$C_{22}H_{30}Cl_2N_{10}$	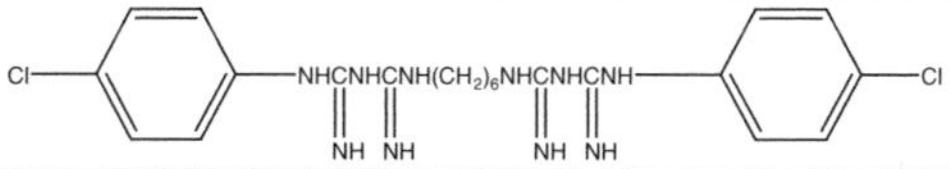 Cl– –NHCNHCNH(CH$_2$)$_6$NHCNHCNH– –Cl; NH NH NH NH
Triclosan	$C_{12}H_7Cl_3O_2$	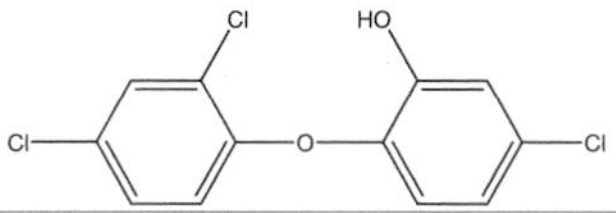 Cl HO Cl –O– Cl
Silver sulfadiazine	$C_{10}H_9AgN_4O_2S$	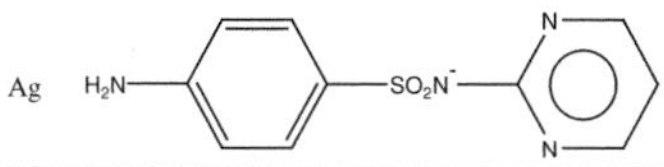 Ag H$_2$N– –SO$_2$N$^-$– N N
Quaternary ammonium salt	$C_{26}H_{58}ClNO_3Si$	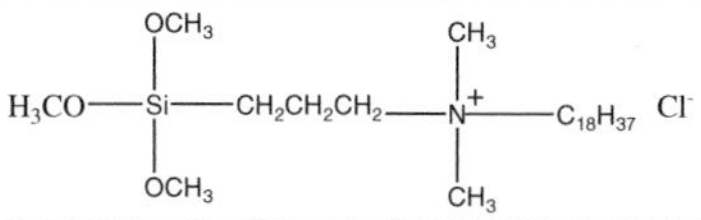 OCH$_3$ CH$_3$ H$_3$CO–Si–CH$_2$CH$_2$CH$_2$–N$^+$–C$_{18}$H$_{37}$ Cl$^-$ OCH$_3$ CH$_3$
Polyhexamethylene biguanide	$(C_8H_{17}N_5)_n$	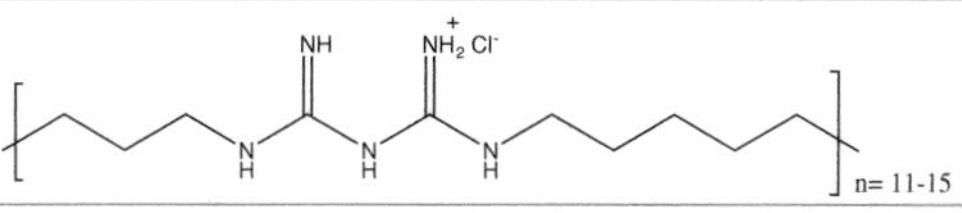 NH NH$_2^+$ Cl$^-$ N H N H N H; n= 11-15

textile should not be negatively altered. Table 1 summarizes some of the commercially available market selling antimicrobial agents.

The most commonly employed synthetic antimicrobial agents for functional finishing are briefly discussed as follows.

2.1.1 Quaternary Ammonium Compounds

These compounds contain positive charge at the nitrogen atom which has been known to be responsible for its antimicrobial activity. The antimicrobial activity also depends upon the length of the alkyl chain, the presence of the per-fluorinated group, and the number of cationic ammonium groups in its chemical structure (Simoncic and Tomsic 2010). QACs can be applied on a wide range of textile materials for effective antimicrobial finishing against a broad spectrum of bacterial and fungal species (Gao and Cranston 2008; Islam et al. 2013a, b). Its mode of action is believed to be by complexing or interaction ability with the negatively charged groups present on microbial cell membranes, thereby altering the cell functions and leading to their death. Though quaternary ammonium compounds provide potential activity to different fibers, it binds via ionic interactions to polyamide fabrics. Attempts have been made to overcome these problems by the use of chemical cross-linking and modifying agents (Simoncic and Tomsic 2010). It is worthy to note that modified quaternary ammonium compounds and composites are now used in order to get bonded to textile surfaces for simultaneously enhancing the antimicrobial property of treated textiles.

2.1.2 *N*-Halamines

N-halamines are chlorinated products of 5,5-dimethylhydantoin and 2,2,5,5-tetramethyl-4-imidazolidinone containing one or two covalent bonds between nitrogen and a halogen atom usually chlorine (Gao and Cranston 2008). Chlorination of amine, amide, or imide groups in dilute sodium hypochlorite can generate nitrogen and chlorine bonds of different stability. *N*-halamines act as biocides for a broad spectrum of bacteria, fungi, and viruses, and this effect is based on the electrophilic substitution of chlorine in the nitrogen–chlorine bond with hydrogen in the presence of water and results in the transfer of Cl^{+} ions that can bind to acceptor regions on microorganisms and hinders enzymatic and metabolic processes, leading to the destruction of the microorganisms (Sun et al. 1995; Worley et al. 1988). *N*-halamines were modified or many compounds were incorporated with them to get high durability to washing.

2.1.3 Polybiguanides

Polybiguanides are polymeric polycationic amines that include cationic biguanide repeat units separated by hydrocarbon chain linkers. Polyhexamethylene biguanide is widely used in medicine as an antiseptic agent, mainly used for preventing wound infection by antibiotic-resistant bacteria (Moore and Gray 2007; Mulder et al. 2007). Their remarkable biocidal activity and low toxicity have attracted attention for antimicrobial finishing of textiles mainly for the protection of cellulose fibers (Blackburn et al. 2007; Kawabata and Taylor 2007; Krebs et al. 2005). Polyhexamethylene biguanide can bind to the anionic carboxylic groups of cellulose through oxidation of glucose rings during pretreatment by bleaching and mercerizing.

2.1.4 Triclosan

It is a halogenated compound commonly known as triclosan 5-chloro-2-(2.4-dichlorophenoxy) phenol and has been used as a biocide in detergents and household objects, including textiles and plastics. Triclosan has been tested as antibacterial agent on cotton against *Escherichia coli* and *Staphylococcus aureus*, and simultaneously exposed to acidic, basic, and synthetic urine solutions (Orhan et al. 2007). It blocks lipid biosynthesis and thereby inhibits bacterial and fungal growth (Gao and Cranston 2008). To get the high durability to washing and to avoid leaching, triclosan is applied to fabrics in the presence of a number of cross-linking agents such as polycarboxylic acids (Guo et al. 2013; Yazdankhah et al. 2006).

2.1.5 Synthetic Dyes

Synthetic dyes were first discovered by W.H. Perkin in 1856 and are now used in different application sectors including dyeing, paper printing, color photography, pharmaceutical, food, cosmetic, and leather industries (Ali 2010). Over the past few decades, more structurally diverse dyes such as acidic, basic, disperse, azo, diazo, anthraquinone, and metal complex dyes have been explored and used for coloration purposes (Islam et al. 2013a, b; Khan et al. 2011). Among all industries, synthetic colorants are more extensively used in textile industry because of their large shade range and good affinity toward textile materials. Unfortunately, the presence of some textile and toxic dyes in effluents generated in wet processing has been continuously polluting waters due to the formation of toxic chemical sludge or carcinogenic compounds (Ali 2010). Some of the dyes also require the use of metallic mordants (Shabbir et al. 2016; Shahid et al. 2012; Shahmoradi Ghaheh et al. 2014; Zhang et al. 2014). Iron, copper, tin, chromium salts are some of the commonly employed metal salts. Several physicochemical methods, such as adsorption, chemical oxidation, precipitation, coagulation, filtration, electrolysis, photodegradation, biological, and microbiological methods, are currently at operation for the removal of toxic dyes from wastewaters. The chemical structure of some of the synthetic dyes is shown in Fig. 2.

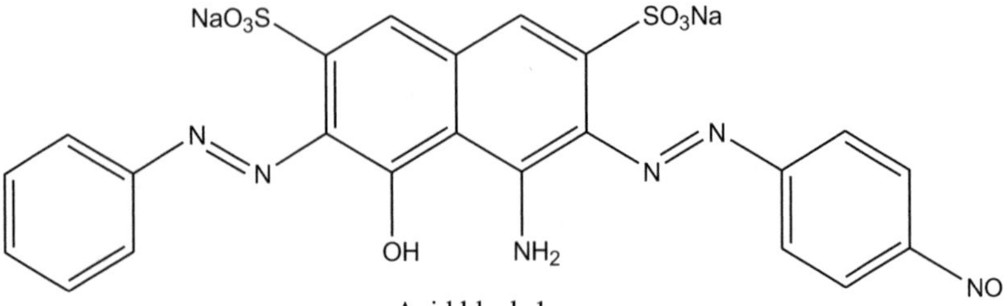

Acid black 1

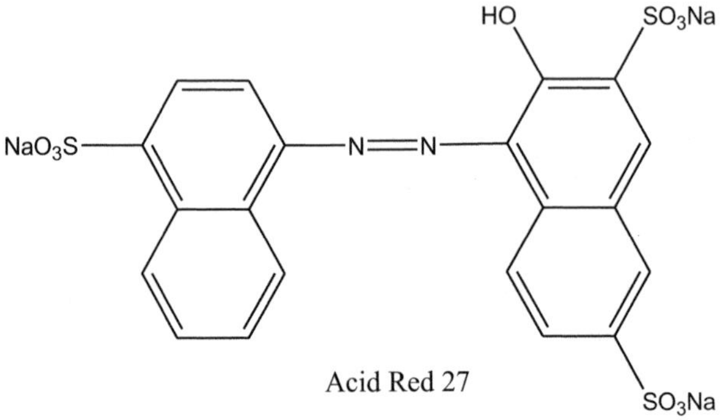

Acid Red 27

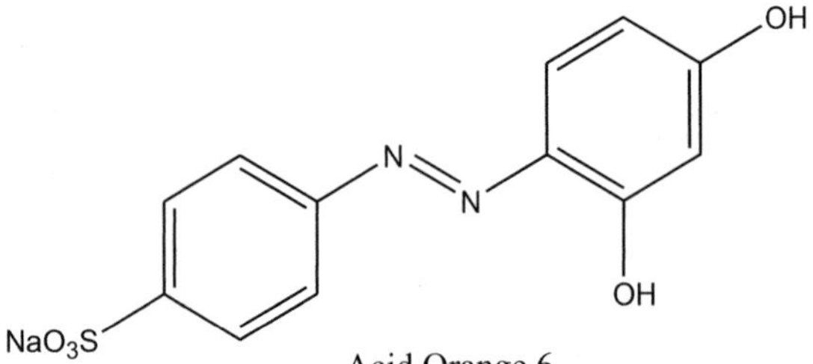

Acid Orange 6

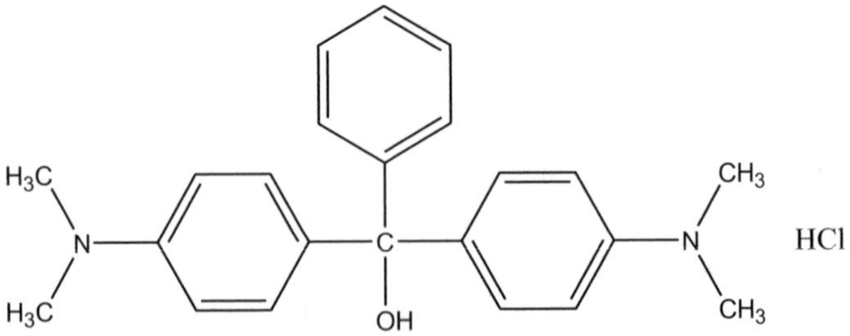

Basic Green 4

Fig. 2 Chemical structures of some synthetic dyes

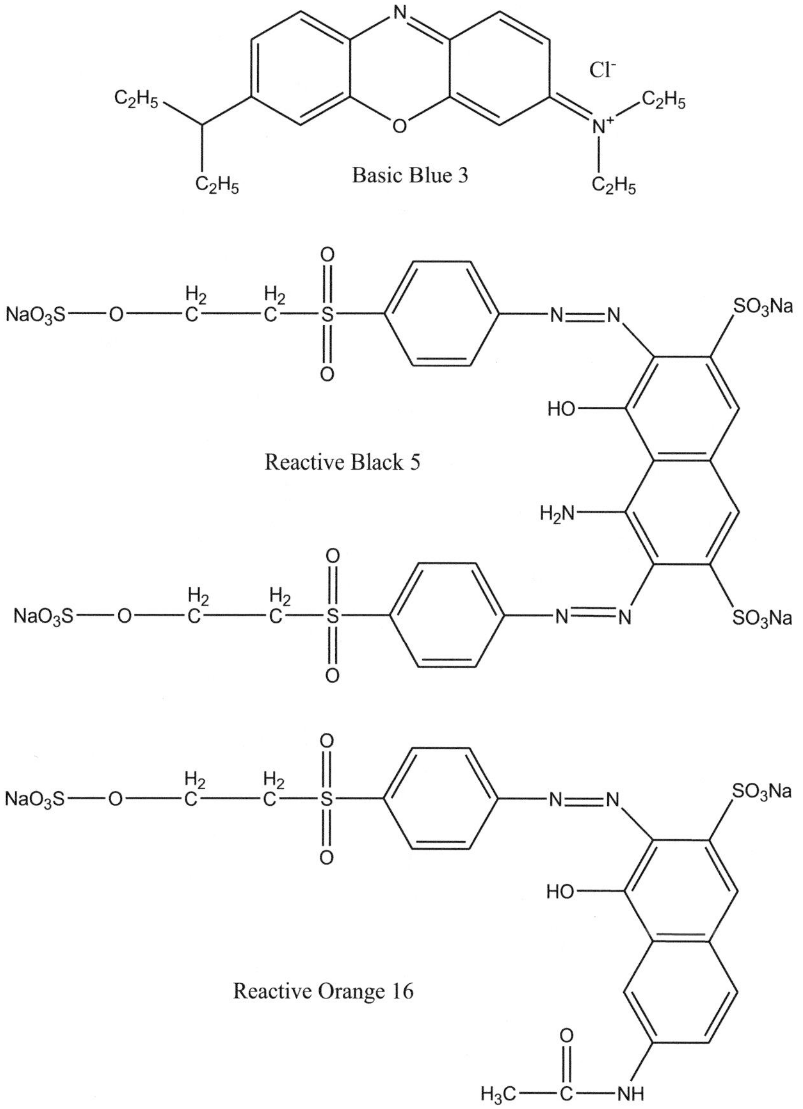

Basic Blue 3

Reactive Black 5

Reactive Orange 16

Fig. 2 (continued)

3 Natural Compounds for Functional Finishing

To mitigate the environmental pollution arising from some synthetic agents, there is a great demand for non-toxic and environmentally friendly natural products for functional finishing of textile materials.

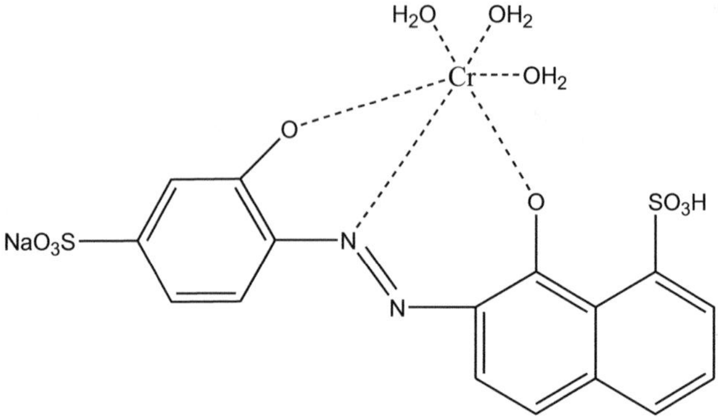

Acid Blue 159 - Metal Complex

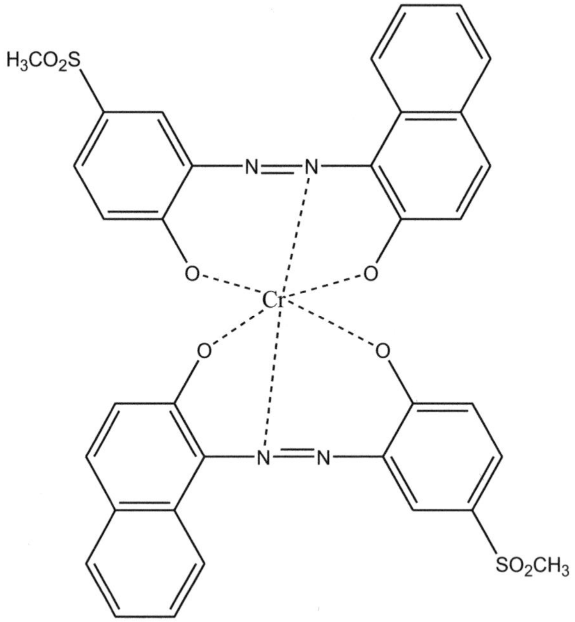

Acid Violet 78 Dye-Metal Complex

Fig. 2 (continued)

3.1 Chitosan

Chitosan is a naturally available biopolymer synthesized from the deacetylation of chitin (Fig. 3). Chitin is found in prawns, crabs, fungi, insects, and other crustaceans (Dev et al. 2009). Being a polysaccharide, chitosan consists of unbranched

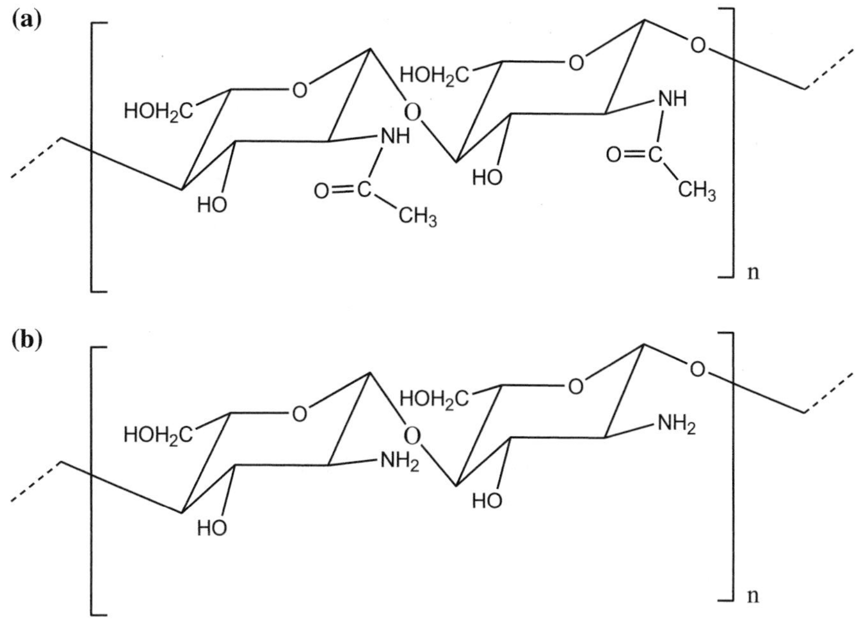

(a)

(b)

Fig. 3 Chemical structure of **a** chitin and **b** chitosan

chains of β-(1 → 4)-2-acetoamido-2-deoxy-D-glucose and is nowadays increasingly being used in functional finishing of textile materials (Islam et al. 2013a, b). Native chitosan has certain limitations such as poor water solubility and chemical reactivity. Researcher has chemically modified chitosan using a number of chemical agents including glyoxal, formaldehyde, glutaraldehyde, epichlorohydrin, ethylene glycol diglycidyl ether and isocyanates to overcome its limitations. Many review papers have been published about its use in textile industry. Griffth reviewed the sources, antimicrobial mechanisms, modifications, dyeing, deodorant, and other functional properties imparted by native and chemically modified chitosan to wool, cotton, silk, and other synthetic textile materials. Apart from this, a recent review article by Islam et al. also highlighted the role of chitosan and its modified derivatives in the development of antimicrobial textiles.

3.2 Plant Extracts

Natural products have been known since prehistoric times for their coloring as well as other functional properties. Till 1856, natural compounds extracted from flora and fauna were the only colorants available for dyeing of clothing and other day-to-day products (Shahid et al. 2012). Over the last few decades, consumers are well aware about ecosystem, and knowledge about environmental pollution has accumulated substantially. Synthetic dyes are often highly toxic and carcinogenic and therefore have been banned by many countries. Several investigations have shown that dyeing wastewaters contain huge amount of toxic and harmful substances and thus have triggered a major concern among scientific community. This has resulted in resurgence of natural dyes once again into dyeing and finishing applications (Vankar et al. 2007, 2008). Natural dyes derived from plant species contain different classes of structurally diverse compounds which produce beautiful hues on wool and cotton (Mirjalili et al. 2011). A number of new dye plants have been identified over the past few decades and have been introduced into textile dyeing applications. Many dyeing compounds isolated from prominent dye plants such as lawsone (2-hydroxy-1,4-naphthoquinone) found in henna (*Lawsonia inermis*) leaves (Yusuf et al. 2012), juglone from *Juglans regia* (Tutak and Korkmaz 2012), tannins from *Quercus infectoria*, and *Punica granatum*, bixin from *Bixa orellana* (Islam et al. 2014), quercetin from onion peel (Adeel et al. 2009), acacetin from *Acacia nilotica* (Rather et al. 2015), and butrin from *Butea monosperma* (Sinha et al. 2012) have resulted in colorful textiles with additional functional characteristics such as antimicrobial and UV protection properties. A more detailed description of natural dyes is covered by a number of review articles published by different groups over the past few years (Islam et al. 2013a, b; Shahid et al. 2013). Other than natural dyes, many herbal products such as aloe vera, tea tree oil, tulsi leaf, azuki beans, and eucalyptus oil have been used in textile applications (Joshi et al. 2009).

Conventional agents generally suffer with serious problems such as less binding efficiency to textile surfaces, and other properties of textile substrates can also be altered with their application such as comfort nature of clothing and durability. Conventional methods as described above often do not produce permanent functionality imparted to textiles surfaces due to which fabrics lose their properties during different processing stages such as wearing, washing, dry-cleaning, and hot pressing. Nanotechnology has the potential to overcome these limitations of conventional methods and provide durability of textile functions (Gao and Cranston 2008; Lombi et al. 2014).

4 Nanomaterials in Functional Finishing

Substantial investigations have revealed the production and application of diverse nanoparticles which include silver, gold, titanium, copper, and zinc onto textile surfaces. Some of the most studied nanomaterials are shown in Fig. 4. Owing to their remarkable properties, they have been employed to fabricate clothing for warmth, comfort, hygienic applications, and style (Dastjerdi et al. 2009). Nanomaterials in various forms such as metal nanoparticles, metal oxides, and nanocomposites are being used for UV protection, conductive, water repellent, antibacterial, and deodorizing functionalization of textiles. They can be applied onto textile surfaces by different means like in situ synthesis of them, spraying and

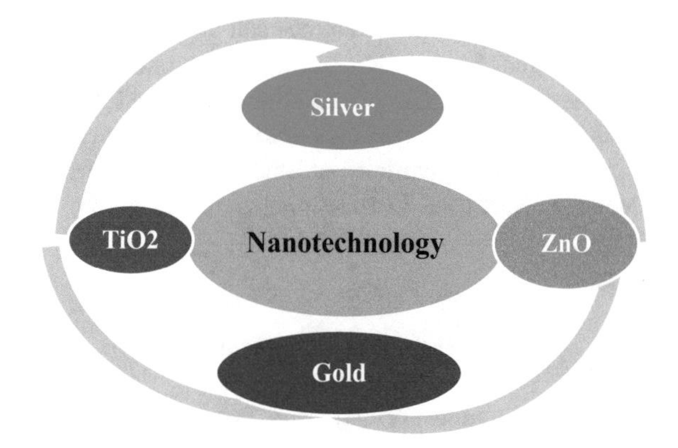

Fig. 4 Commonly used nanomaterials in functional finishing of textile materials

other wet processes. To get the deposition coefficient of high degree, textiles can be treated first with other means also such as plasma.

4.1 Overview of Commonly Employed Nanoparticles in Functional Finishing

Physical and chemical methods are the most popular methods for the production of nanoparticles. Over the past few decades, production of metal nanoparticles by green chemistry method using plant extracts and enzymes is receiving much attention as some chemical methods cannot avoid the use of toxic and hazardous reagents therefore posing environmental challenges (Islam et al. 2014). Plants are enriched by a variety of natural products including reducing and stabilizing agents and hence seem to be the best candidates for large-scale biosynthesis of different nanoparticles (Iravani 2011). The synthesized nanoparticles find application in different fields including textile dyeing and finishing. Silver nanoparticles are the most studied nanomaterials for functionalization of textiles. They have strong antimicrobial activity and produce different hues depending up on their size and shape. A number of reports are available in the literature on the application of silver nanoparticles onto cotton, wool, silk, polyamide, and other synthetic textile surfaces (Dubas et al. 2006; Rai et al. 2009). Because of their surface plasma resonance property, silver nanoparticles have been employed to impart yellow, brown, red, gray, and other novel shades on textile surfaces. Both in situ and ex situ methods of silver nanoparticles application have been tested, and it has been discovered that in situ is advantageous method to functionalize different textile products Perera et al. (2013).

Apart from silver, gold nanoparticles are also very important for application on wool and cotton textiles. Different sized gold nanoparticles are used to produce a range of colors on natural and synthetic textiles. Au nanogold rods were recently applied on cotton to produce UV protection and color. Several cross-linkers such as monomeric or polymeric amines can be used to obtain stable gold nanoparticles by reducing the Au^{+3} to Au^{o}. TEM, X-ray photoelectron spectroscopy, and electron microscopy are some of the commonly used techniques to characterize the formation of gold nanoparticles onto textile surfaces (Johnston et al. 2008).

Likewise, nanoparticles of other metals such as Cu, Ni, Fe, and Co can also be deposited and used in functionalization of textile materials (Vigneshwaran et al. 2009).

4.1.1 Metal Oxides and Composites

Titanium dioxide (TiO_2) due to its extraordinary photocatalytic activity, non-toxicity, high availability, biocompatibility, and low price is attracting scientific attention for use in different application fields (Dastjerdi and Montazer 2010). Titanium dioxide has resulted in the production of antibacterial, UV protective, and self-cleaning properties on textile materials. A number of review articles have been published on the textile applications of TiO_2 nanoparticles. Montazer and Pakdel reviewed the use of TiO_2 in functional finishing of textile materials with a more emphasis on wool (Montazer and Pakdel 2011). Apart from this review, Radetic (2013) published a more comprehensive review collecting dispersed information on antibacterial activity, UV protection, self-cleaning finishing of cotton, wool, and other synthetic textile substrates with $TiO_{2.}$

Likewise, zinc oxide nanoparticles posses remarkable antibacterial, UV-blocking, superhydrophobic, photocatalyst activities. They have created new remarkable approach as multifunctional finishing agent for textile materials Perelshtein et al. (2009). The different methods of synthesis, characterization, and application of zinc oxide onto textile substrates are recently review by Montazer et al. (2013).

5 Applications

5.1 Antibacterial Textiles

Textile materials including wool and silk are suitable substrate for hosting odor-generating bacteria, fungi, and moulds (Islam et al. 2016). These textiles provide suitable environment such as moisture, oxygen, nutrients, and temperature for multiplication of drug-resistant pathogens (Dastjerdi and Montazer 2010). Consumer's enhanced awareness about general sanitation, cross-transmission of diseases, and personal protection has motivated researchers to develop antibacterial and antifungal clothing (Khan et al. 2011). To get antimicrobial functionalization, nanomaterials are nowadays the subject of considerable interest. Nanoparticles offer several advantages such high durability and less resistance, and are more stable over

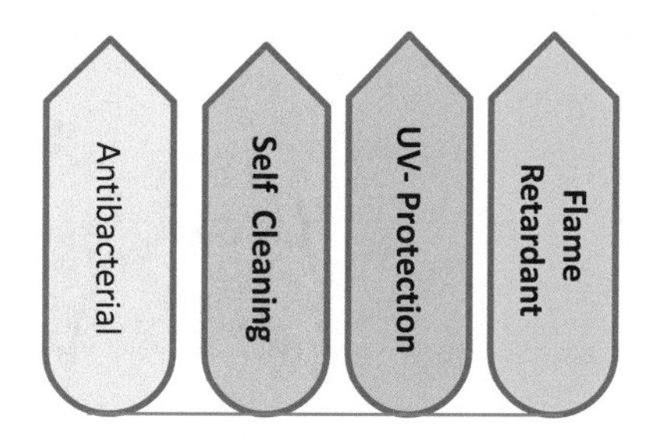

Fig. 5 Some important functionality imported by nanoparticles

organic biocides for antimicrobial application (Perelshtein et al. 2009; Sawai 2003). Figure 5 depicts some of the important functionalities imparted to textile materials.

5.2 UV Protection Clothing

Ultraviolet rays mainly UVC (<290 nm) radiations reaching the earth from sun are considered as harmful for skin, and their continuous exposure causes many skin problems such as sunburns, premature aging, allergies, and skin cancers. This has necessitated the development of UV protective clothing. Textile material, its construction, and kind of finishing agent used are the main factors that determine the UV property (Saravanan 2007). Among several UV-blocking agents, TiO_2 nanoparticles incorporated into fabrics during spinning have resulted in the protection of textiles against photoyellowing and photooxidation impacts of UV-C rays. To increase the stability of TiO_2 nanoparticles on textile surfaces, many cross-linking agents have been used. Likewise, zinc oxide nanoparticles are versatile, and potential UV-blocking functional finishing agents researched round the globe. Zinc oxide has been utilized to impart UV-blocking property to wool, cotton, silk, and other textile surfaces (Kathirvelu et al. 2009). Two recent review articles published by a group led by Montazer have highlighted the UV-blocking property of TiO_2 and zinc oxide nanoparticles in detail, respectively.

5.3 Self-cleaning Textiles

Self-cleaning surface is highly desired in many advanced application fields. Different kinds of impurities, dirt molecules, including stains with tea, coffee, body perspiration, are mainly responsible for unhygienic nature of textile (Montazer et al. 2011). These impurities can be removed by hydrophobic and hydrophilic surfaces (Fig. 6). Nanotechnology is booming and playing extraordinary role. Titanium dioxide is a promising photocatalyst and has been extensively used in self-cleaning apparel over the past few years. Meilert et al. (2005) made efforts to enhance binding TiO_2 on cotton surfaces using chemical spacers. Bozzi et al. (2005) used

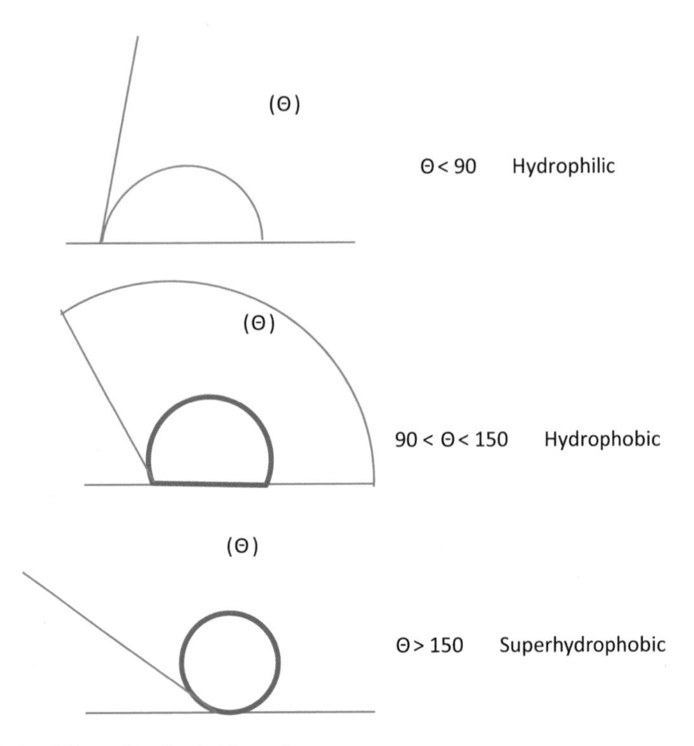

Fig. 6 Hydrophilic and hydrophobic surfaces

radio frequency plasma (RF-plasma), microwave plasma (MW-plasma), and vacuum-UV light irradiation as pretreatments to modify textile material in order to enhance discoloration of stains contained in wine and coffee. A more detailed description about self-cleaning textiles is presented in a recent review article published by Radetic (2013).

5.4 Flame-Retardant Textiles

All textile materials such as cellulose, protein, and synthetic polymers such as polyesters, polyamides, polyacrylonitrile, and cellulose acetate provide ideal conditions and pose serious fire hazards in case of fire accidents for property and personal loses (Neisius et al. 2015). In the present scenario, nanotechnology is a emerging technology and is a promising source for flame-retardant agents which have been widely used in flame-retardant textiles. The nanoparticles may act as barrier to limit the heat, fuel, and oxygen. Flame retardancy can also be achieved with application of nanostructures such as nanoparticles of metals, metal oxides, and nanocomposites with permanent effect on textiles. Bourbigot et al. reported the synthesis of polyamide-6/clay nanocomposite using melt blending and investigated its flammability. The produced nanocomposite polyamide-6/clay was found to impart permanent effect at relatively low cost and without altering basic properties of textiles (Bourbigot et al. 2002). In another investigation conducted by Federico Carosio et al. 2011 it was shown that silica nanoparticles could be applied on textile by layer-by-layer assembly to increase time to ignition and decreased heat release rate peak of fabric.

6 Current Challenges and Perspectives

The customers' demand for textiles with multifunctional properties has motivated researchers to introduce nanotechnology into textile modification applications. As already discussed, nanoparticles are materials in 1–100 nm size range having high surface area-to-volume ratio. Nanotechnology revolutionized the textile sector over the past few decades (Montazer et al. 2013; Dastjerdi and Montazer 2010). Many textile fibers such as wool, cotton, silk, and synthetic fibers such as polyester and nylon are widely used in many advanced application fields. It is well described in the literature that many nanoparticles such as titanium dioxide and zinc oxide can be applied to impart resistance, water repellency, ultraviolet protection, and electromagnetic and infrared functional finishing properties to textile materials (Montazer and Pakdel 2011). Other nanoparticles have been utilized to induce antibacterial, antifungal, and wrinkle resistance properties to textile surfaces. Nanotechnology modification provides a new way to impart novel prominent properties to textile surfaces without affecting other bulk properties. Since the textile industries are biggest consumer base, it is expected that nanotechnology would play a major role in its economy. Other than their largest use in apparel market, nanotextiles could find potential use in different application fields including hospital textiles, sportswear, military, domestic, and household furnishing (Radetic 2013).

It may be noted that though nanotechnology has fascinated textile industries, little attention has been paid to its impact on environment and human health. The release of nanoparticles should be monitored carefully during washing. Several research investigations have been carried out in this area, and it has been established that the forms of nanomaterials released depend upon the way they are incorporated, the characteristics of liquid, pH, and other attributes (Lombi et al. 2014). Therefore, the future research should be targeted to study in detail the ways to reduce or control the pollution and toxicity imposed by nanoparticles (Schilling et al. 2010; Windler et al. 2012). By exploiting the greener ways for nanoparticles production and their textile applications, nanotechnology offers full potential for commercialization. Consumer's enhanced awareness to reduce environmental pollution has already started and forced textile and clothing manufactures and regulatory agencies to ensure that there is less toxicity during production and life cycle of nanoparticles. Therefore, in the years to come, we hope that nanotechnology will be a boom in every area of textile research.

7 Conclusion

The nanotechnology has been a robust technology and offers many advantages in textile finishing. As compared to conventional methods, nanoparticles in view of large surface area-to-volume ratio produce durable functions into textile materials. A wide range of nanoparticles including silver, gold, titanium dioxide, copper, zinc oxide, and aluminum oxide have been described to impart high tensile strength, wettability, hydrophobicity, and to produce some novel properties such as soft hand, durability, water repellency, fire retardancy, and antimicrobial properties. In the near future, it is expected that nanoproducts will be manufactured commercially provided there is more and more research regarding their costs and ways to reduce their uncontrolled release into wastewaters causing environmental pollution and human health risks.

References

Adeel, S., Ali, S., Bhatti, I. A., & Zsila, F. (2009). Dyeing of cotton fabric using pomegranate (*Punica granatum*) aqueous extract. *Asian Journal of Chemistry, 21*(5), 3493.

Ali, H. (2010). Biodegradation of synthetic dyes—A review. *Water, Air, and Soil pollution, 213* (1–4), 251–273.

Anghel, I., Grumezescu, A. M., Andronescu, E., Anghel, A. G., Ficai, A., Saviuc, C., et al. (2012). Magnetite nanoparticles for functionalized textile dressing to prevent fungal biofilms development. *Nanoscale research letters, 7*(1), 2–7.

Aniołczyk, H., Koprowska, J., Mamrot, P., & Lichawska, J. (2004). Application of electrically conductive textiles as electromagnetic shields in physiotherapy. *Fibres and Textiles in Eastern Europe, 12*(4), 47–50.

Batool, F., Adeel, S., Azeem, M., Khan, A. A., Bhatti, I. A., Ghaffar, A., et al. (2013). Gamma radiations induced improvement in dyeing properties and colorfastness of cotton fabrics dyed with chicken gizzard leaves extracts. *Radiation Physics and Chemistry, 89*, 33–37.

Bhatti, I. A., Adeel, S., Jamal, M. A., Safdar, M., & Abbas, M. (2010). Influence of gamma radiation on the colour strength and fastness properties of fabric using turmeric (*Curcuma longa* L.) as natural dye. *Radiation Physics and Chemistry, 79*(5), 622–625.

Blackburn, R. S., Harvey, A., Kettle, L. L., Manian, A. P., Payne, J. D., & Russell, S. J. (2007). Sorption of chlorhexidine on cellulose: Mechanism of binding and molecular recognition. *Journal of Physical Chemistry B, 111*(30), 8775–8784.

Bourbigot, S., Devaux, E., & Flambard, X. (2002). Flammability of polyamide-6/clay hybrid nanocomposite textiles. *Polymer Degradation and Stability, 75*(2), 397–402.

Bozzi, A., Yuranova, T., & Kiwi, J. (2005). Self-cleaning of wool-polyamide and polyester textiles by TiO_2-rutile modification under daylight irradiation at ambient temperature. *Journal of Photochemistry and Photobiology A: Chemistry, 172*, 27–34.

Carosio, F., Laufer, G., Alongi, J., Camino, G. & Grunlan, J. C. (2011). Layer-by-layer assembly of silica-based flame retardant thin film on PET fabric. *Polymer Degradation and Stability, 96* (5), 745–750.

Dastjerdi, R., & Montazer, M. (2010). A review on the application of inorganic nano-structured materials in the modification of textiles: Focus on anti-microbial properties. *Colloids and Surfaces B: Biointerfaces, 79*(1), 5–18.

Dastjerdi, R., Montazer, M., & Shahsavan, S. (2009). A new method to stabilize nanoparticles on textile surfaces. *Colloids and Surfaces A: Physicochemical and Engineering Aspects, 345*(1), 202–210.

Dev, V. G., Venugopal, J., Sudha, S., Deepika, G., & Ramakrishna, S. (2009). Dyeing and antimicrobial characteristics of chitosan treated wool fabrics with henna dye. *Carbohydrate Polymers, 75*(4), 646–650.

Dubas, S. T., Kumlangdudsana, P., & Potiyaraj, P. (2006). Layer-by-layer deposition of antimicrobial silver nanoparticles on textile fibers. *Colloids and Surfaces A: Physicochemical and Engineering Aspects, 289*(1), 105–109.

Gao, Y. & Cranston, R. (2008). Recent advances in antimicrobial treatments of textiles. *Textile Research Journal, 78*(1), 60–72.

Guo, C., Zhou, L., & Lv, J. (2013). Effects of expandable graphite and modified ammonium polyphosphate on the flame-retardant and mechanical properties of wood flour-polypropylene composites. *Polymers and Polymer Composites, 21*(7), 449–456.

Hakansson, E., Kaynak, A., Lin, T., Nahavandi, S., Jones, T., & Hu, E. (2004). Characterization of conducting polymer coated synthetic fabrics for heat generation. *Synthetic Metals, 144*(1), 21–28.

Hutchison, J. E. (2008). Greener nanoscience: A proactive approach to advancing applications and reducing implications of nanotechnology. *ACS Nano, 2*(3), 395–402.

Iravani, S. (2011). Green synthesis of metal nanoparticles using plants. *Green Chemistry, 13*, 2638–2650.

Islam, S., Butola, B. S., & Mohammad, F. (2016). Silver nanomaterials as future colorants and potential antimicrobial agents for natural and synthetic textile materials. *RSC Advances, 6*, 44232–44247.

Islam, S., & Mohammad, F. (2014). Emerging green technologies and environment friendly products for sustainable textiles. In: S. S. Muthu (ed.), *Roadmap to sustainable textiles and clothing* (pp. 6–82): Singapore: Springer.

Islam, S., Shahid, M., & Mohammad, F. (2013a). Green chemistry approaches to develop antimicrobial textiles based on sustainable biopolymers—A review. *Industrial and Engineering Chemistry Research, 52*, 5245–5260.

Islam, S., Shahid, M., & Mohammad, F. (2013b). Perspectives for natural product based agents derived from industrial plants in textile applications—A review. *Journal of Cleaner Production, 57*, 2–18.

Islam, S., Shahid, M., & Mohammad, F. (2014). Future prospects of phytosynthesized transition metal nanoparticles as novel functional agents for textiles. In: A. T. Syväjärvi (ed.), *Advanced materials for agriculture, food, and environmental safety* (pp. 265–290). New York: Wiley.

Johnston, J. H., Richardson, M. J., & Burridge, K. A. (2008). *Gold nanoparticles as colourants in high fashion fabrics and textiles* (Vol. 1, pp. 712–715).

Joshi, M., Wazed Ali, S., Purwar, R., & Rajendran, S. (2009). Ecofriendly antimicrobial finishing of textiles using bioactive agents based on natural products. *Indian Journal of Fibre & Textile Research, 34*(3), 295–304.

Kale, K. H., Palaskar, S. S., & Kasliwal, P. M. (2012). *A novel approach for functionalization of polyester and cotton textiles with continuous online deposition of plasma polymers, 37* (September), 238–244.

Kaplan, S., & Okur, A. (2008). The meaning and importance of clothing comfort: A case study for Turkey. *Journal of Sensory Studies, 23*(5), 688–706.

Kathirvelu, S., D'souza, L., & Dhurai, B. (2009). UV protection finishing of textiles using ZnO nanoparticles. *Indian Journal of Fibre Textile Research*, *34*(3), 267–273

Kawabata, A., & Taylor, J. A. (2007). The effect of reactive dyes upon the uptake and antibacterial efficacy of poly(hexamethylene biguanide) on cotton. Part 3: Reduction in the antibacterial efficacy of poly(hexamethylene biguanide) on cotton, dyed with bis(monochlorotriazinyl) reactive dyes. *Carbohydrate Polymers, 67*(3), 375–389.

Khan, M. I., Ahmad, A., Khan, S. A., Yusuf, M., Shahid, M., Manzoor, N., et al. (2011). Assessment of antimicrobial activity of catechu and its dyed substrate. *Journal of Cleaner Production, 19*(12), 1385–1394.

Khan, S. A., Ahmad, A., Khan, M. I., Yusuf, M., Shahid, M., Manzoor, N., et al. (2012). Antimicrobial activity of wool yarn dyed with *Rheum emodi* L. (Indian Rhubarb). *Dyes and Pigments, 95*(2), 206–214.

Krebs, F. C., Miller, S. R., Ferguson, M. L., Labib, M., Rando, R. F., & Wigdahl, B. (2005). Polybiguanides, particularly polyethylene hexamethylene biguanide, have activity against human immunodeficiency virus type 1. *Biomedicine & Pharmacotherapy, 59*(8), 438–445.

Li, D., & Sun, G. (2007). Coloration of textiles with self-dispersible carbon black nanoparticles. *Dyes and Pigments, 72*(2), 144–149.

Lombi, E., Donner, E., Scheckel, K. G., Sekine, R., Lorenz, C., Goetz, N. V. et al. (2014). Silver speciation and release in commercial antimicrobial textiles as influenced by washing. *Chemosphere, 111*, 352–358.

Meilert, K. T., Laub, D., & Kiwi, J. (2005). Photocatalytic self-cleaning of modified cotton textiles by TiO_2 clusters attached by chemical spacers. *Journal of Molecular Catalysis A: Chemical, 237*, 101–108.

Mirjalili, M., Nazarpoor, K., & Karimi, L. (2011). Eco-friendly dyeing of wool using natural dye from weld as co-partner with synthetic dye. *Journal of Cleaner Production, 19*(9), 1045–1051.

Montazer, M., Amiri, M. M. & Malek, R. M. A. (2013). In situ synthesis and characterization of nano ZnO on wool: influence of nano photo reactor on wool properties. *Photochemistry and photobiology, 89*(5), 1057–1063.

Montazer, M., & Maali Amiri, M. (2014). ZnO nano reactor on textiles and polymers: Ex situ and in situ synthesis, application, and characterization. *The Journal of Physical Chemistry B, 118* (6), 1453–1470.

Montazer, M., & Pakdel, E. (2011). Functionality of nano titanium dioxide on textiles with future aspects: Focus on wool. *Journal of Photochemistry and Photobiology C: Photochemistry Reviews, 12*(4), 293–303.

Montazer, M., Pakdel, E., & Behzadnia, A. (2011). Novel feature of nano-titanium dioxide on textiles: Antifelting and antibacterial wool. *Journal of Applied Polymer Science, 121*(6), 3407–3413.

Moore, K., & Gray, D. (2007). Using PHMB antimicrobial to prevent wound infection. *Wounds uK, 3*(2), 96–102.

Mulder, G. D., Cavorsi, J. P., & Lee, D. K. (2007). Polyhexamethylene niguanide (PHMB): An addendum to current topical antimicrobials. *Wounds : A Compendium of Clinical Research and Practice, 19*(7), 173—182. http://europepmc.org/abstract/MED/26110333

Neisius, M., Stelzig, T., Liang, S., & Gaan, S. (2015). *Flame retardant finishes for textiles*. Functional Finishes for Textiles: Woodhead Publishing Limited. doi:10.1533/9780857098450.2.429

Orhan, M., Kut, D., & Gunesoglu, C. (2007). Use of triclosan as antibacterial agent in textiles. *Indian Journal of Fibre & Textile Research, 32*, 114–118.

Perelshtein, I., Applerot, G., Perkas, N., Wehrschetz-Sigl, E., Hasmann, A., Guebitz, G. M., et al. (2009). Antibacterial properties of an in situ generated and simultaneously deposited nanocrystalline ZnO on fabrics. *ACS Applied Materials and Interfaces, 1*(2), 361–366

Perera, S., Bhushan, B., Bandara, R., & Rajapakse, G. (2013). Colloids and surfaces A: Physicochemical and engineering aspects morphological, antimicrobial, durability, and physical properties of untreated and treated textiles using silver-nanoparticles. *Colloids and Surfaces A: Physicochemical and Engineering Aspects, 436*, 975–989.

Petkova, P., Francesko, A., Fernandes, M. M., Mendoza, E., Perelshtein, I., Gedanken, A., et al. (2014). Sonochemical coating of textiles with hybrid ZnO/chitosan antimicrobial nanoparticles.

Radetić, M. (2013). Functionalization of textile materials with TiO_2 nanoparticles. *Journal of Photochemistry and Photobiology C: Photochemistry Reviews, 16*, 62–76.

Rai, M., Yadav, A., & Gade, A. (2009). Silver nanoparticles as a new generation of antimicrobials. *Biotechnology Advances, 27*(1), 76–83.

Rather, L. J., Islam, S. & Mohammad, F. (2015). Study on the application of *Acacia nilotica* natural dye to wool using fluorescence and FT-IR spectroscopy. *Fibers and Polymers, 16*(7), 1497–1505.

Saravanan, D. (2007). UV protection textile materials. *Autex Research Journal, 7*(1), 53–62.

Sawai, J. (2003). Quantitative evaluation of antibacterial activities of metallic oxide powders (ZnO, MgO and CaO) by conductimetric assay. *Journal of Microbiological Methods, 54*(2), 177–182.

Schilling, K., Bradford, B., Castelli, D., Dufour, E., Nash, J. F., Pape, W., et al. (2010). Human safety review of "nano" titanium dioxide and zinc oxide. *Photochemical & Photobiological Sciences, 9*(4), 495–509.

Shabbir, M., Islam, S. U., Bukhari, M. N., Rather, L. J., Khan, M. A., & Mohammad, F. (2016). Application of *Terminalia chebula* natural dye on wool fiber—Evaluation of color and fastness properties. *Textiles and Clothing Sustainability, 2*(1), 1.

Shahid, M., Ahmad, A., Yusuf, M., Khan, M. I., Khan, S. A., Manzoor, N., et al. (2012). Dyeing, fastness and antimicrobial properties of woolen yarns dyed with gallnut (*Quercus infectoria* Oliv.) extract. *Dyes and Pigments, 95*(1), 53–61.

Shahid, M., Islam, S., & Mohammad, F. (2013). Recent advancements in natural dye applications: a review. *Journal of Cleaner Production, 53*, 310–331.

Shahmoradi Ghaheh, F., Mortazavi, S. M., Alihosseini, F., Fassihi, A., Shams Nateri, A., & Abedi, D. (2014). Assessment of antibacterial activity of wool fabrics dyed with natural dyes. *Journal of Cleaner Production, 72*, 139–145.

Simoncic, B., & Tomsic, B. (2010). Structures of novel antimicrobial agents for textiles—A review. *Textile Research Journal, 80*(16), 1721–1737.

Sinha, K., Saha, P. D., & Datta, S. (2012). Extraction of natural dye from petals of flame of forest (*Butea monosperma*) flower: Process optimization using response surface methodology (RSM). *Dyes and Pigments, 94*(2), 212–216.

Sun, G., Chen, T. Y., Sun, W., Wheatley, W. B. & Worley, S. D. (1995). Preparation of novel biocidal N-halamine polymers. *Journal of bioactive and compatible polymers, 10*(2), 135–144.

Textile-Based Drug Release Systems. (n.d.). doi:10.1533/9781845692933.1.50

Tutak, M., & Korkmaz, N. E. (2012). Environmentally friendly natural dyeing of organic cotton. *Journal of Natural Fibers, 9*(1), 51–59.

Vankar, P. S., Shanker, R., Mahanta, D., & Tiwari, S. C. (2008). Ecofriendly sonicator dyeing of cotton with *Rubia cordifolia* Linn. using biomordant. *Dyes and Pigments, 76*(1), 207–212.

Vankar, P. S., Shanker, R., & Verma, A. (2007). Enzymatic natural dyeing of cotton and silk fabrics without metal mordants. *Journal of Cleaner Production, 15*(15), 1441–1450.

Vigneshwaran, N., Varadarajan, P. V, & Balasubramanya, R. H. (2009). Application of metallic nanoparticles in textiles. *Nanotechnologies for the Life Sciences*, 541–558.

Windler, L., Lorenz, C., Von Goetz, N., Hungerbuhler, K., Amberg, M., Heuberger, M., et al. (2012). Release of titanium dioxide from textiles during washing. *Environmental Science and Technology, 46*(15), 8181–8188.

Worley, S. D., Williams, D. E., & Crawford, R. A. (1988). Halamine water disinfectants. *Critical Reviews in Environmental Control, 18*(2), 133–175.

Yazdankhah, S. P., Scheie, A., Høiby, E. A., Lunestad, B.-T., Heir, E., Fotland, T. Ø., et al. (2006). Triclosan and antimicrobial resistance in bacteria: An overview. *Microbial Drug Resistance (Larchmont, N.Y.), 12*(2), 83–90.

Yuranova, T., Rincon, A. G., Pulgarin, C., Laub, D., Xantopoulos, N., Mathieu, H. J., et al. (2006). Performance and characterization of Ag-cotton and Ag/TiO_2 loaded textiles during the abatement of *E. coli*. *Journal of Photochemistry and Photobiology A: Chemistry, 181*(2–3), 363–369.

Yusuf, M., Ahmad, A., Shahid, M., Khan, M. I., Khan, S. A., Manzoor, N., et al. (2012). Assessment of colorimetric, antibacterial and antifungal properties of woollen yarn dyed with the extract of the leaves of henna (*Lawsonia inermis*). *Journal of Cleaner Production, 27*, 42–50.

Zhang, B., Wang, L., Luo, L., & King, M. W. (2014). Natural dye extracted from Chinese gall—The application of color and antibacterial activity to wool fabric. *Journal of Cleaner Production, 80*, 204–210.

Printed in the United States
By Bookmasters